Construction Health and Safety in Coastal and Maritime Engineering

Construction Health and Safety in Coastal and Maritime Engineering

Ravinder Bora

RANDOM PUBLICATIONS
NEW DELHI (INDIA)

Construction Health and Safety in Coastal and Maritime Engineering

ISBN 978-93-5111-234-1

Published in 2014 in India by

RANDOM PUBLICATIONS

4376-A/4B, Gali Murari Lal, Ansari Road
New Delhi-110 002
Phone : +91-11-43580356, +91-11-23289044
e-mail: randomexports@gmail.com, sales@randompublications.com, info@randompublications.com

Reprinted 2026

Type Setting by : Keystoneprintads, Delhi-110051

Preface

Over US$ 700 million is spent on coastal and maritime construction. Just under half of this expenditure relates to commercial projects including ports, harbors, docks, and jetties, with the other half being spent on coastal and flood defense work. Coastal and maritime construction work is particularly hazardous due to the hostile environment in which work is carried out.

The volatility and unpredictability of conditions caused by storms, waves, currents, and tides have resulted in a number of major accidents and fatalities in recent years. At present, there is little specific health and safety related guidance available to assist clients, designers, planning supervisors, contractors and other stakeholders to ensure that work in this sector is undertaken in a safe manner. This book has been produced following detailed consultation with a wide range of design engineers, clients, and contractors. It covers best practice in safety management of coastal and maritime design and construction work.

Best practice information on safe working practices in the coastal and maritime engineering sector has been assembled through examination of the Health and Safety Executive (HSE) statistical data and through detailed discussions with clients, consultants, contractors, and suppliers, examining current operational procedures, accident records, and perceptions of risk. This book incorporates lessons learnt from author's experience in the coastal and maritime construction sectors, and contains guidance and recommendations for all stakeholders involved in the safety of construction work in the coastal and maritime environment.

The Construction (Design and Management) Regulations (CDM) Regulations, among other regulations, require both the designer and contractor to consider and assess construction risks. Without specific information and guidance on safe practice in the marine environment, it is not only difficult to undertake safety assessments but these assessments may be misguided and incorrect. The aim of this book is to provide stakeholders with specific guidance on how to avoid, combat at source, mitigate, and protect personnel from risks in coastal and maritime engineering works.

I thank all members of my team who have helped in the preparation of the book. My special thanks go to "Random Publications" who have published the book.

– ***Ravinder Bora***

Contents

1

Marine Vehicle

OVERVIEW

Seagoing marine vehicles may be divided into *transport,* including cargo, container and passenger ships and *nontransport* including fishing vessels, service craft such as tugs and supply vessels, and warships. Each vessel type has a particular role to play and each will have a different set of design and operational conditions. This chapter provides a brief review of the main design and operational features of the principal types of marine vehicle.

MERCHANT SHIPS

The development of merchant ship types has been dictated largely by the nature of the cargo and the trade routes. They can be classified accordingly with the major types being:

- General cargo ships
- Container ships
- Tankers
- Dry bulk carriers
- Passenger ships
- Tugs.

GENERAL CARGO SHIPS

The industry distinguishes between *break bulk* cargo which is packed, loaded and stowed separately and *bulk* cargo which is carried loose in bulk. The general cargo carrier is a flexible design of vessel which will go anywhere and carry a wide variety of cargo. The cargo may be break bulk or containers. Such vessels have several large clear open cargo-carrying spaces or holds. One or more decks may be present within the holds. These are known as ' tween decks and provide increased flexibility in loading and unloading, permit cargo segregation and improved stability. Access to the holds is by openings in the deck known as hatches.

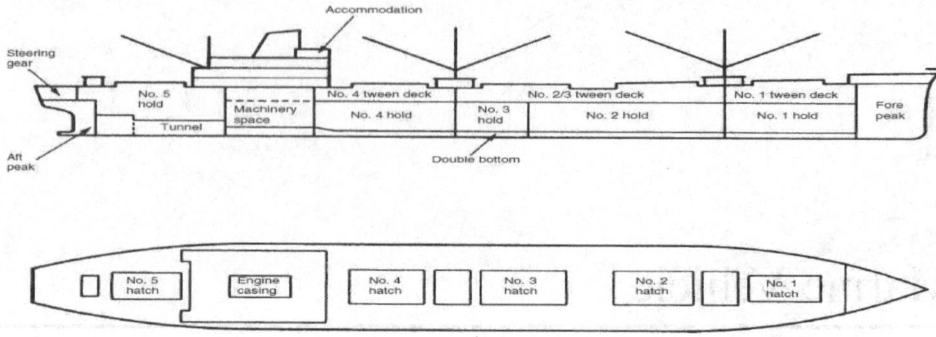

Fig. General cargo ship

Hatches are made as large as strength considerations permit in order to reduce the amount of horizontal movement of cargo within the ship. Typically the hatch width is about a third of the ship's beam. Hatch covers are of various types. Pontoon hatches are quite common in ships of up to 10 000 dwt, for the upper deck and ' tween decks, each pontoon weighing up to 25 tonnes. They are opened and closed using a gantry or cranes. In large bulk carriers side rolling hatch covers are often fitted, opening and closing by movement in the transverse direction. Another type of cover is the folding design operated by hydraulics. T

he coamings of the upper or weather deck hatches are raised above the deck to reduce the risk of flooding in heavy seas. They are liable to distort a little due to movement of the structure during loading and unloading of the ship. This must be allowed for in the design of the securing arrangements. Coamings can provide some compensation for the loss of hull strength due to the deck opening. A double bottom is fitted along the ship's length, divided into various tanks. These may be used for fuel, lubricating oils, fresh water or ballast water.

Fore and aft peak tanks are fitted and may be used to carry ballast and to trim the ship. Deep tanks are often fitted and used to carry liquid cargoes or water ballast. Water ballast tanks can be filled when the ship is only partially loaded in order to provide a sufficient draught for stability, better weight distribution for longitudinal strength and better propeller immersion. Cranes and derricks are provided for cargo handling. Typically cranes have a lifting capacity of 10–25 tonnes with a reach of 10–20 m, but they can be much larger. General cargo ships can carry cranes or gantries with lifts of up to 150 tonnes. Above this, up to about 500 tonnes lift they are referred to as heavy lift ships. The machinery spaces are often well aft but there is usually one hold aft of the accommodation and machinery space to improve the trim of the vessel when partially loaded. General cargo ships are generally smaller than the ships devoted to the carriage of bulk cargoes. Typically their speeds range from 12 to 18 knots.

Refrigerated cargo ships (Reefers)

A refrigeration system provides low temperature holds for carrying perishable cargoes. The holds are insulated to reduce heat transfer. The cargo may be carried frozen or chilled and holds are at different temperatures according to requirements. The possible effect of the low temperatures on surrounding structure must be considered. Refrigerated fruit is carried under modified atmosphere conditions.

The cargo is maintained in a nitrogen-rich environment in order to slow the ripening process. The costs of keeping the cargo refrigerated, and the nature of the cargo, make a shorter journey time desirable and economic and these vessels are usually faster than general cargo ships with speeds up to 22 knots. Up to 12 passengers are carried on some, this number being the maximum permitted without the need to meet full passenger ship regulations.

Container ships

Container ships are a good example of an integrated approach to the problem of transporting goods. Once goods are placed in the container at a factory or depot, they can be carried by road, rail or sea, being transferred from one to another at road or rail depots or a port.

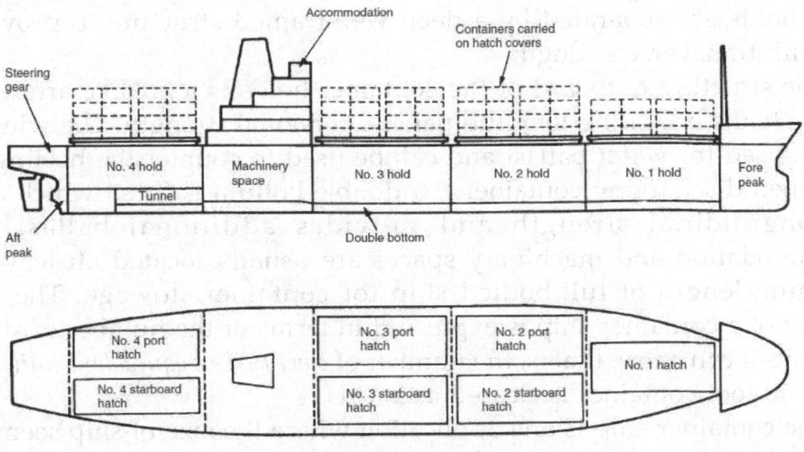

Fig. Container ship

The container need not be opened until it reaches its destination. This makes the operation more secure. The maritime interest is primarily in the ports and ships but any element of the overall system may impose restrictions on what can be done. Height of container is likely to be dictated by the tunnels and bridges involved in land transport. Weight is likely to be dictated by the wheel loadings of lorries. The handling arrangements at the main terminals and ports are specially designed to handle the containers quickly and accurately. The larger container ships use dedicated container ports and tend not to have their own cargo handling gantries. The containers themselves are simply reusable boxes made of steel or aluminium. They come in a range of types and sizes. Details can be obtained from the web site of one of the operators. Nominal dimensions are lengths of 20, 40 and 45 ft, width of 8 ft and height 8.5 or 9.5 ft. Internal volumes and weight of goods that can be carried vary with the material. For a 20 ft general-purpose steel container the internal capacity is about 33 m 3, weight empty is about 2.3 tonnef and the maximum payload is about 21.7 tonnef.

Figures for a 40 ft container would be about 68 m 3, 3.8 tonnef and 26.7 tonnef, respectively. Aluminium containers will have about the same volume, weigh less and be able to carry a larger payload. They are used for most general cargoes and liquid carrying. The cargo-carrying section of the ship is divided into several holds with the containers racked in special frameworks and stacked one upon the other within the hold space.

Containers may also be stacked on hatch covers and secured by special lashings. Some modern ships dispense with the hatch covers, pumps dealing with any water that enters the holds. Each container must be of known all up weight and stowage arrangements must ensure the ship's stability is adequate as well as meeting the offloading schedule if more than one port is involved. The ship's deadweight will determine the total number of containers carried. Cargo holds are separated by a deep web-framed structure to provide the ship with transverse strength.

The structure outboard of the container holds is a boxlike arrangement of wing tanks providing longitudinal and torsional strength. The wing tanks may be used for water ballast and can be used to counter the heeling of the ship when discharging containers. A double bottom is fitted which adds to the longitudinal strength and provides additional ballast space. Accommodation and machinery spaces are usually located aft leaving the maximum length of full-bodied ship for container stowage. The overall capacity of a container ship is expressed in terms of the number of standard 20 ft units it can carry, that is, the number of *twenty-foot equivalent units* (TEU). Thus a 40-foot container is classed as 2 TEU.

The container ship is one application where the size of ship seems to be ever increasing to take advantage of the economies of scale. By the turn of the century 6000 TEU ships had become the standard for the main trade routes, and some 80 ships of 8000 TEU were on order plus some of 9200 TEU. Concept work was underway for ships of 14 000 TEU size. Container ships tend to be faster than most general cargo ships, with speeds up to 30 knots. The larger ships can use only the largest ports. Since these are fitted out to unload and load containers the ship itself does not need such handling gear. Smaller ships are used on routes for which the large ships would be uneconomic, and to distribute containers from the large ports to smaller ports. That is, they can be used as feeder ships. Since the smaller ports may not have suitable handling gear the ships can load and offload their own cargos. Some containers are refrigerated. They may have their own independent cooling plant or be supplied with cooled air from the ship's refrigeration system. Because of the insulation required refrigerated containers have less usable volume. Temperatures would be maintained at about –27°C and for a freezer unit about –60°C. They may be carried on general cargo ships or on dedicated refrigerated container ships. One such dedicated vessel is a 21 knot, 30 560 dwt ship of 2046 TEU capacity.

The ship has six holds of which five are open. The hatchcover-less design enables the cell structure, in which the containers are stowed, to be continued above deck level giving greater security to the upper containers. Another advantage of the open hold is the easier dissipation of heat from the concentration of reefer boxes. Barge carriers are a variant of the container ship. Standard barges are carried into which the cargo has been previously loaded. The barges, once unloaded, are towed away by tugs and return cargo barges are loaded. Minimal or even no port facilities are required and the system is particularly suited to countries with extensive inland waterways.

Roll-on roll-off ships (Ro-Ro ships)

These vessels are designed for wheeled cargo, often in the form of trailers. The cargo can be rapidly loaded and unloaded through stern or bow doors and sometimes sideports for smaller vehicles. Train ferries were an early example of Ro-Ro ships. The cargo-carrying section of the ship is a large open deck with a loading ramp usually at the aft end. Internal ramps lead from the loading deck to the other ' tween deck spaces. The cargo may be driven aboard under its own power or loaded by straddle carriers or fork lift trucks. One or more hatches may be provided for containers or general cargo, served by deck cranes. Where cargo, with or without wheels, is loaded and discharged by cranes the term lift-on lift-off (Lo-Lo) is used.

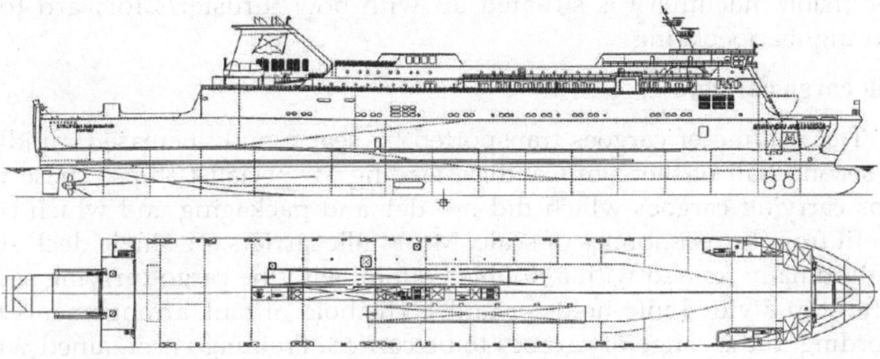

Fig. Ro-Ro ship

The structure outboard of the cargo decks is a boxlike arrangement of wing tanks to provide longitudinal strength and adequate transverse stability. A double bottom is fitted along the complete length. Transverse bulkheads are limited to below the lowest vehicle deck so the side structure must provide adequate transverse and torsional strength. The machinery space and accommodation are located aft. Only a narrow machinery casing actually penetrates the loading deck. Sizes range considerably with about 16 000 dwt (28 000 displacement tonne) being quite common and speeds are relatively high and usually in the region of 18–22 knots. The use of Ro-Ro ships as passenger ferries is discussed later.

Car carriers

The increasing volume of car and truck production in the East (Japan, Korea and China) and a large customer base in the West has seen the introduction and rapid increase in number of ships specifically designed and built to facilitate the delivery of these vehicles globally. Probably the ugliest ships afloat, car carriers are strictly functional having a very high box-like form above the waterline to accommodate as many vehicles as possible on, in some cases as many as a dozen decks.

Whilst most deck spacing is to suit cars some ' tween deck heights may be greater and the deck strengthened to permit loading of higher and heavier vehicles. Within such greater deck spacing lift able car decks may be fitted for flexibility of stowage. The spacing of fixed car decks can vary from 1.85 to 2.3 metres to accommodate varying shape and height of cars. Transfer arrangements for vehicles from the main deck are by means of hoistable ramps which can be lifted and lowered whilst bearing the vehicles. Loading and discharging vehicles onto and off the ship is via a large quarter ramp at the stern and a side shell or stern ramp.

The crew accommodation and forward wheelhouse, providing an adequate view forward, sit atop the uppermost continuous weather deck. Propulsion machinery is situated aft with bow thruster/s forward to aid mooring/manoeuvring.

Bulk cargo carriers

The volume of cargoes transported by sea in bulk increased rapidly in the second half of the 20th Century, leading to specialist ships. These were ships carrying cargoes which did not demand packaging and which could benefit from the economies of scale. Most bulk carriers are single deck ships, longitudinally framed with a double bottom, with the cargo-carrying section of the ship divided into holds or tanks. The hold or tank arrangements vary according to the range of cargoes to be carried. Framing is contained within the double bottom and wing tanks to leave the inner surfaces of the holds smooth. They are categorized as:

- *Panamax.* The dimensions of the ship being limited by the need to be able to transit the Panama Canal. The beam must be less than 32.25 m.
- *Suezmax.* The dimensions of the ship being limited by the need to be able to transit the Suez Canal. Draught to be less than 19 m.
- *Capesize.* Without the restrictions of the above types.
- *Handysize.* Generally less than about 50 000 tonnes.
- *Aframax.* This is a term applied to tankers in the range 80 000–120 000 dwt.

Bulk carriers can also be sub-divided into tankers and dry bulk carriers. The requirements, for instance the permitted lengths of cargo holds, vary with the size of ship and the following comments are for general guidance only.

Tankers

Tankers are used for the transport of liquids. They nclude:

- Crude oil carriers;
- Product tankers;
- Gas tankers;
- Chemical carriers.

Crude oil carriers

These carry the unrefined crude oil and they have significantly increased in size in order to obtain the economies of scale and to respond to the demands for more and more oil. Designations, such as Ultra Large Crude Carrier (ULCC) and Very Large Crude Carrier (VLCC), have been used for these huge vessels. The ULCC is a ship of 300 000 dwt or more; the VLCC is 200 000–300 000 dwt. Crude oil tankers with deadweight tonnages in excess of half a million have been built although the current trend is for somewhat smaller (130 000–150 000 dwt) vessels.

The cargo-carrying section of the tanker is usually divided into tanks by longitudinal and transverse bulkheads. The size and location of these cargo tanks is dictated by the International Maritime Organization (IMO) Convention MARPOL 1973/1978 which became internationally accepted in 1983. IMO requirements are built into those of the various classification societies.

These regulations require the use of segregated ballast tanks and their location such that they provide a barrier against accidental oil spillage. The segregated ballast tanks must be such that the vessel can operate safely in ballast without using any cargo tank for water ballast. See Chapter 11 for more information on IMO and MARPOL. Tankers ordered after 1993 had to comply with the MARPOL double hull regulation.

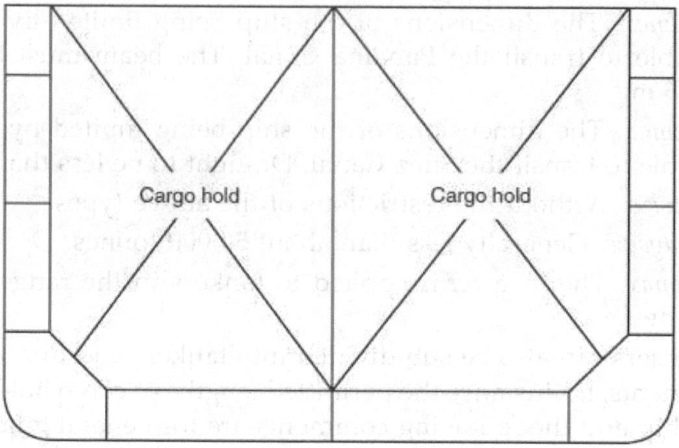

Fig. Typical section of double hull tanker

This is opposed to single hull tankers where one or more cargo holds are bounded in part by the ship's shell plating. In the double hull design the cargo tanks are completely surrounded by wing and double bottom tanks which can be used for ballast purposes. The USA, under its 1990 Oil Pollution Act required all newly built tankers trading in US waters to be of the double hull design. There has been debate on whether a double hull is the best way of reducing pollution following grounding or collision. IMO and classification societies are prepared to consider alternatives to the double hull. One alternative favoured by some is the mid-height depth deck design. In such ships a deck is placed at about mid-depth which will be well below the loaded waterline.

This divides the cargo tanks into upper and lower tanks. A trunk is taken from the lower tank through the upper tank and vented. The idea is that if the outer bottom is breached the external water pressure will be greater than the pressure of hull from the lower tank and this will force oil up the vent trunk. Thus water enters the ship rather than oil escaping from it. Such tankers would still incorporate segregated ballast tanks outboard of the cargo tanks to safeguard against collision. Subsequent debate within IMO and the EU has led to a speeding up of the timetable for phasing out single hull tankers. For the detailed provisions recourse should be had to the regulations of the authorities concerned.

Segregated ballast tanks would include all the double bottom tanks beneath the cargo tanks, wing tanks and the fore and aft peak tanks. Each cargo tank would be discharged by pumps fitted in the aft pump room, each tank having its own suction arrangement which connects to the pumps, and a network of piping discharges the cargo to the deck from where it is pumped ashore. The accommodation and machinery spaces would be located aft and separated from the tank region by a cofferdam. Where piping serves several tanks, means must be provided for isolating each tank. Experience shows that once any initial protective coatings breakdown, permanent ballast tanks suffer corrosion and regular inspection is vital. The builder must provide a high quality coating system and a back-up anode system to give a coverage of 10 mA/m 2 should be included to control corrosion after coating breakdown. More and more ships are being fitted with equipment to measure actual strains during service. A typical system comprises a number of strain gauges at key points in the structure together with an accelerometer and pressure transducer to monitor bottom impacts. Results are available on the bridge to assist the master in the running of the ship. The information is stored and is invaluable in determining service loadings and long term fatigue data.

Product carriers

After the crude oil is refined the various products are transported in product carriers. The refined products carried include gas oil, aviation fuel and kerosene. Product carriers are smaller than crude oil carriers and, because several different products are carried, they have greater sub-division of tanks. The cargo tank arrangement is again dictated by MARPOL 73/78. Individual ' parcels ' of various products may be carried at any one time which results in several separate loading and discharging pipe systems. The tank surfaces are usually coated to prevent contamination and enable a high standard of tank cleanliness to be achieved after discharge. Sizes range from about 18 000 up to 75 000 dwt with speeds of about 14–16 knots.

Liquefied gas carriers

The most commonly carried liquefied gases are *liquefied natural gas* (LNG) and *liquefied petroleum gas* (LPG). They are kept in liquid form by a combination of pressure and low temperature. The combination varies to suit the gas being carried. The bulk transport of natural gases in liquefied form began in 1959 and has steadily increased since then. Specialist ships are used to carry the different gases in a variety of tank systems, combined with arrangements for pressurizing and refrigerating the gas. Natural gas is released as a result of oil-drilling operations. It is a mixture of methane, ethane, propane, butane and pentane. The heavier gases, propane and butane, are termed ' petroleum gases '. The remainder, largely methane, is known as ' natural gas '. The properties, and behaviour, of these two basic groups vary considerably, requiring different means of containment and storage during transit.

Natural gas carriers

Natural gas is, by proportion, 75–95% methane and has a boiling point of –162°C at atmospheric pressure. Methane has a critical temperature of –82°C, which means it cannot be liquefied by the application of pressure above this temperature. A pressure of 47 bar is necessary to liquefy methane at –

82°C. LNG carriers are designed to carry the gas in its liquid form at atmospheric pressure and a temperature in the region of –164°C. The ship design must protect the steel structure from the low temperatures, reduce the loss of gas and avoid its leakage into the occupied regions of the vessel, Ffooks. Tank designs are either self-supporting, membrane or semi-membrane. The self-supporting tank is constructed to accept any loads imposed by the cargo. A membrane tank requires the insulation between the tank and the hull to be load bearing. Single or double metallic membranes may be used, with insulation separating the two membrane skins. The semi-membrane design has an almost rectangular cross section and the tank is unsupported at the corners.

The tank and insulation structure is surrounded by a double hull. The double bottom and ship's side regions are used for oil or water ballast tanks whilst the ends provide cofferdams between the cargo tanks. A pipe column is located at the centre of each tank and is used to route the pipes from the submerged cargo pumps out of the tank through the dome. The accommodation and machinery spaces are located aft and separated from the tank region by a cofferdam. LNG carriers have steadily increased in size and ships of around 140 000 m^3 capacity are now on order. Speeds range from 16 to 19 knots.

Petroleum gas carriers

Petroleum gas may be propane, propylene, butane or a mixture. All three have critical temperatures above normal ambient temperatures and can be liquefied at low temperatures at atmospheric pressure, normal temperatures under considerable pressure, or some intermediate combination of pressure and temperature. The design must protect the steel hull where low temperatures are used, reduce the gas loss, avoid gas leakage and perhaps incorporate pressurized tanks. The fully pressurized tank operates at about 17 bar and is usually spherical or cylindrical in shape for structural efficiency. Semi-pressurized tanks operate at a pressure of about 8 bar and temperatures in the region of _ 7°C.

Insulation is required and a reliquefaction plant is needed for the cargo boil-off. Cylindrical tanks are usual and may penetrate the deck. Fully refrigerated atmospheric pressure tank designs may be selfsupporting, membrane or semi-membrane types as in LNG tankers. The fully refrigerated tank designs operate at temperatures of about –45°C. A double hull type of construction is used.

Within the holds there are four independent, insulated, prismatic cargo tanks, supported by a load bearing structure designed to take account of the interaction of movements and forces between the tanks and adjoining hull members. Topside wing, hopper side and double bottom tanks are mainly used for water ballast. Fuel is carried in a cross bunker forward of the engine room. Machinery and accommodation are right aft. It can carry various propane/butane ratios to provide flexibility of operation.

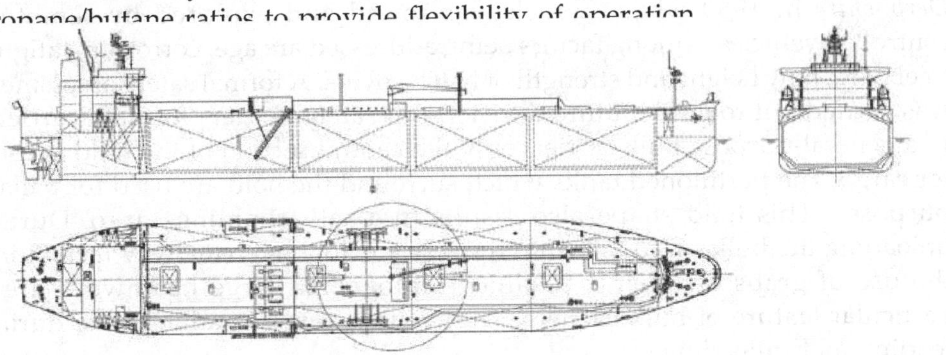

Fig. LPG carrier

The double hull construction, cargo pumping arrangements, accommodation and machinery location are similar to an LNG carrier. A reliquefaction plant is, however, carried and any cargo boil-off is returned to the tanks. LPG carriers exist in sizes up to around 95 000 m^3. Speeds range from 16 to 19 knots.

Chemical carriers

A wide variety of chemicals is carried by sea. The cargo is often toxic and flammable so the ships are subject to stringent requirements to ensure safety of the ship and the environment. Different cargos are segregated by cofferdams.

Spaces are provided between the cargo tanks and the ship's hull, machinery spaces and the forepeak bulkhead. Great care is taken to prevent fumes spreading to manned spaces.

Dry bulk carriers

These ships carry bulk cargoes such as grain, coal, iron ore, bauxite, phosphate and nitrate. Towards the end of the 20th Century more than 1000 million tonnes of these cargoes were being shipped annually, including 180 million tonnes of grain. Apart from saving the costs of packaging, loading and offloading times are reduced. As the volume of cargo carried increased so did the size of ship, taking advantage of improving technology. By the 1970s ships of 200 000 dwt were operating and even larger ships were built later.

This growth in size has not been without its problems. In the 28 months from January 1990 there were 43 serious bulk carrier casualties of which half were total losses. Three ships, each of over 120 000 dwt, went missing. Nearly 300 lives were lost as a result of these casualties. To improve the safety of these ships IMO adopted a series of measures during the 1990s. These reflect the lessons learned from the losses of ships in the early 1990s and that of *MV Derbyshire* in 1980 whose wreck was found and explored by remotely controlled vehicles. Among factors being addressed are age, corrosion, fatigue, freeboard, bow height and strength of hatch covers. A formal safety assessment was carried out to guide future decisions on safety matters for bulk carriers. In a general-purpose bulk carrier, only the central section of the hold is used for cargo. The partitioned tanks which surround the hold are used for ballast purposes. This hold shape also results in a self-trimming cargo. During unloading the bulk cargo falls into the space below the hatchway facilitating the use of grabs or other mechanical unloaders. Large hatchways are a particular feature of bulk carriers. They reduce cargo handling time during loading and unloading.

Combination carriers are bulk carriers which have been designed to carry any one of several bulk cargoes on a particular voyage, for instance ore, crude oil or dry bulk cargo. Stability and loading manuals are supplied to every ship to provide the Master with the information to load, discharge and operate the ship safely. Loading computer programs are designed to provide, for any condition of loading, a full set of deadweight, trim, stability and longitudinal strength calculations.

IMO codes on trimming bulk cargoes require the cargo, with particular attention to cargoes that may liquefy, to be trimmed reasonable level to the boundaries of the compartment to minimize the risk of bulk material shift. The very high loading rates, up to 16 000 tonnes/hour, make the loading task one that needs careful attention. An ore carrier usually has two longitudinal bulkheads which divide the cargo section into wing tanks and a centre hold which is used for ore.

A deep double bottom is fitted. Ore, being a dense cargo, would have a very low centre of gravity if placed in the hold of a normal ship leading to an excess of stability in the fully loaded condition. The deep double bottom raises the centre of gravity and the behaviour of the vessel at sea is improved. The wing tanks and the double bottoms provide ballast capacity. An ore/oil carrier uses two longitudinal bulkheads to divide the cargo section in centre and wing tanks which are used for the carriage of oil. When ore is carried, only the centre tank section is used for cargo. A double bottom is fitted but used only for water ballast.

The ore/bulk/oil (OBO) bulk carrier is currently the most popular combination bulk carrier. It has a cargo-carrying cross section similar to the general bulk carrier but the structure is significantly stronger. Large hatches facilitate rapid cargo handling. Many bulk carriers do not carry cargo handling equipment, since they trade between special terminals with special equipment. Combination carriers handling oil cargoes have their own cargo pumps and piping systems for discharging oil. They are required to conform to the requirement of MARPOL. Deadweight capacities range from small to upwards of 200 000 tonnes. Taking a 150 000/160 000 tonne deadweight Capesize bulk carrier as typical, the ship is about 280 m in length, 45 m beam and 24 m in depth. Nine holds hold some 180 000 m^3 grain in total, with ballast tanks of 75 000 m^3 capacity. The speed is about 15.5 knots on 14 MW power. Accommodation about 30.

Passenger ships

Passenger ships can be considered in two categories, the cruise ship and the ferry. The ferry provides a link in a transport system and often has Ro-Ro facilities in addition to its passengers. Considerable thought has been given to achieving rapid, and safe, evacuation and this is an area where computer simulation has proved very useful.

For instance, quicker access is possible to lifeboats stowed lower in the ship's superstructure, chutes or slides can be used for passengers to enter lifeboats already in the water, either directly into the boat or by using a transfer platform. It is important that such systems should be effective in adverse weather conditions and when the ship is heeled. Shipboard arrangements must be designed bearing in mind the land-based rescue organizations covering the areas in which the ship is to operate. Free fall lifeboats, used for some years on offshore installations, are increasingly being fitted to tankers and bulk carriers.

Drop heights of 30 m are now accepted and heights of 45 m have been tested. However, safe usage depends upon the potential users being fit and well trained. These conditions can be met in ships ' crews but is problematic for passenger ships. Passengers in, say, a cruise ship may not be fit and may even be partially handicapped. As might be expected it is passenger ships that are most affected by changes in standards and thinking of society as a whole. In 1997 the Maritime and Coastguard agency issued a guidance note on the needs of disabled people. In 2000 the Ferries Working Group of the Disabled Persons Transport Advisory Committee (DPTAC) issued more detailed guidance, DPTAC.

Cruise ships

Cruise ships have been a growth area. Between 1990 and 2000 the cruise market grew by 60% and the size of ship has also grown with vessels now capable of carrying 3600 passengers at 22 knots. However, the largest cruise ships cannot use some ports and harbours in the more attractive locations. The ship has to anchor well out and ferry passengers ashore by smaller boats. This takes time and there are now a number of small or medium sized ships to cater for passengers who want to visit the smaller islands. In a cruise ship passengers are provided with a high standard of accommodation and leisure facilities. This results in a large superstructure as a prominent feature of the vessel. The many tiers of decks are fitted with large open lounges, ballrooms, swimming pools and promenade areas. Stabilizers are fitted to reduce rolling and bow thrusters are used to improve manoeuvrability.

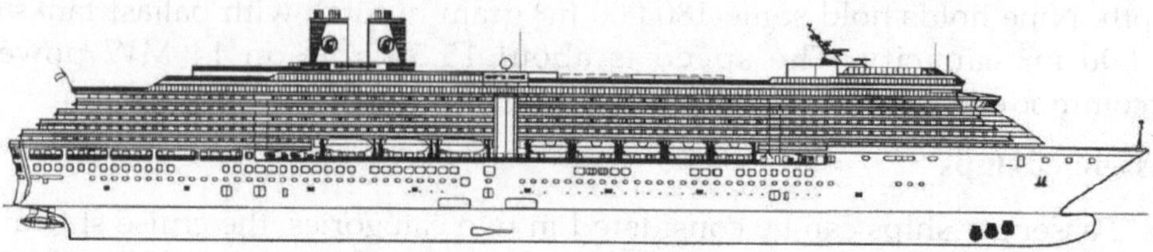

Fig. Cruise liner

Large ferries

Ocean-going ferries are a combination of Ro-Ro and passenger vessel. The vessel has three layers, the lower machinery space, the vehicle decks and the passenger accommodation. A large stern door and sometimes also a bow door provide access for the wheeled cargo to the various decks which are connected by ramps. Great care is needed to ensure these doors are watertight and proof against severe weather. There is usually a secondary closure arrangement in case the main door should leak. The passenger accommodation varies with length of the journey. For short-haul or channel crossings public rooms with aircraft-type seats are provided and for long distance ferries cabins and sleeping berths.

Stabilizers and bow thrusters are usually fitted to improve seakeeping and manoeuvring. Size varies according to route requirements and speeds are usually around 20–22 knots. When used as ferries, vehicles usually enter at one end and leave at the other. This speeds up loading and unloading but requires two sets of doors. There has been considerable debate on the vulnerability of Ro-Ro ships that should water get on to their vehicle decks. Various means of improving stability in the event of collision and to cater for human error in not securing entry doors, have been proposed. Since the loss of the *Herald of Free Enterprise* regulations have been tightened up. The later loss of the *Estonia* gave an additional impetus to a programme of much needed improvements.

Tugs

Tugs perform a variety of tasks and their design varies accordingly. They move dumb barges, help large ships manoeuvre in confined waters, tow vessels on ocean voyages and are used in salvage and firefighting operations. Tugs can be categorized broadly as inland, coastal or ocean going. Put simply, a tug is a means of applying an external force to any vessel it is assisting. That force may be applied in the direct or the indirect mode. In the former the major component of the pull is provided by the tug's propulsion system. In the latter most of the pull is provided by the lift generated by the flow of water around the tug's hull, the tug's own thrusters being mainly employed in maintaining its attitude in the water. The main features of a tug are an efficient design for free running and a high thrust at zero speed (the *bollard pull*), an ability to get close alongside other vessels, good manoeuvrability and stability.

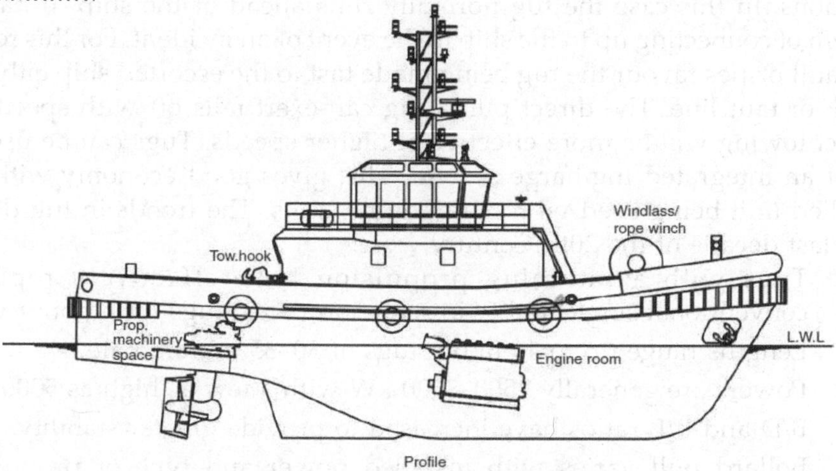

Fig. Tug

Another way of classifying tugs is by the type and position of the propulsor units:

- *Conventional tugs* have a normal hull, propulsion being by shafts and propellers, which may be open or nozzled, and of fixed or controllable pitch, or by steerable nozzles or vertical axis propellers. They usually tow from the stern and push with the bow.
- *Stern drive tugs* have the stern cut away to accommodate twin azimuthing propellers. These propellers, of fixed or controllable pitch, are in nozzles and can be turned independently through 360° for good manoeuvrability. Because the drive is through two right angle drive gears these vessels are sometimes called Z-drive tugs. They usually have their main winch forward and tow over the bow or push with the bow.
- *Tractor tugs* are of unconventional hull form, with propulsors sited under the hull about one-third of the length from the bow, and a stabilizing skeg aft. Propulsion is by azimuthing units or vertical axis propellers. They usually tow over the stern or push with the stern.

In most tug assisted operations the ship is moving at low speed. Concern for the environment, following the *Exxon Valdez* disaster, led to the US Oil Pollution Act of 1990. To help tankers, or any ship carrying hazardous cargo, which found themselves unable to steer, escort tugs were proposed. In this concept the assisted ship may be moving at 10 knots or more. Success depends upon the weather conditions and the proximity of land or underwater hazards, as well as the type and size of tug. Some authorities favour a free-running tug so as not to endanger ship or tug in the majority (incident free) of operations. In this case the tug normally runs ahead of the ship. It has the problem of connecting up to the ship in the event of an incident. For this reason other authorities favour the tug being made fast to the escorted ship either on a slack or taut line. The direct pull a tug can exert falls off with speed and indirect towing will be more effective at higher speeds. Tugs can be used as part of an integrated tug/barge system. This gives good economy with one propelled unit being used with many dumb units. The trends in tug design in the last decade of the 20th Century were:

- Tugs with azimuthing propulsion have effectively replaced conventional single and twin screw propelled tugs for harbour work.
- Lengths range up to 45 m but tugs of 30–35 m dominate.
- Powers are generally 2500–3000 kW with a few as high as 5000 kW.
- B/D and B/L ratios have increased to provide greater stability.
- Bollard pull varies with installed power and type of propulsion, being 60–80 tonnef at 5000 kW.
- Free-running speeds range from 10 to 15 knots and tend to increase linearly with the square root of the length.

Icebreakers and ice strengthened ships

The main function of an icebreaker is to clear a passage through ice at sea, in rivers or in ports so that other ships can use the areas which would otherwise be denied to them. Icebreakers are vital to the economy of nations such as Russia with ports that are ice bound for long periods of the year and which wish to develop the natural resources within the Arctic. Icebreakers need:

- To be specially strengthened with steels which remain tough at low temperature.
- Extra structure in the bow and along the waterline.
- High power propulsion and manoeuvring devices which are not susceptible to ice damage. The shape of the stern is important here.
- A hull form that enables them to ride up over the ice. This is one way of forcing a way through ice; the ship rides over the ice edge and uses its weight to break the ice. The ship may be ' rocked ' by transferring ballast water longitudinally. The hull is well rounded and may roll heavily as protruding stabilizers are unacceptable.
- Good hull sub-division.
- Special hull paints.

Icebreakers are expensive to acquire and operate. Other ships which need to operate in the vicinity of ice are strengthened to a degree depending upon the perceived risk. Usually they can cope with continuous 1 year old ice of 50–100 cm thickness.

Typically they are provided with a double hull, thicker plating forward and in the vicinity of the waterline, with extra framing. They have a flat hull shape and a rounded bow form. Rudders and propellers are protected from ice contact by the hull shape. Inlets for engine cooling water must not be allowed to become blocked.

Fishing vessels

Fishing vessels have evolved over thousands of years to suit local conditions. Fish which live at the bottom of the sea like sole, hake and halibut and those which live near the bottom like cod, haddock and whiting are called demersal species.

Those fish which live above the bottom levels, predominantly such as herring and mackerel, are called pelagic species. There are also three fundamental ways of catching fish:

- By towing trawls or dredges;
- By surrounding the shoals by nets, purse seines;
- By static means, lines, nets or pots.

The commonest type of fishing vessel is the trawler which catches both demersal and pelagic species. The trawl used for the bottom is long and stocking shaped and is dragged at a few knots by cables led to the forward gantry on the ship. When the trawl is brought up it releases its catch in the cod end down the fish hatch in the trawl deck. Operations are similar when trawling for pelagic species but the trawl itself has a wider mouth and is altogether larger.

Trawlers suffer the worst of weather and are the subject of special provision in the freeboard regulations. They must be equipped with machinery of the utmost reliability since failure at a critical moment could endanger the ship. Both diesel and dieselelectric propulsion are now common. Ice accretion in the upperworks is a danger in certain weather, and a minimum value of GM of about 0.75 m is usually required by the owner. Good range of stability is also important and broaching to is an especial hazard. Despite great improvements in trawler design significant numbers of vessels are lost every year and many of them disappear without any very good explanation.

It is probable that such losses are due to the coincidence of two or more circumstances like broaching to, open hatches, choked freeing ports, loss of power, critical stability conditions, etc. To give adequate directional stability when trawling, experience has shown that considerable stern trim is needed, often as much as 5 degrees. Assistance in finding shoals of fish is given by sonar or echo sounding gear installed in the keel. No modern trawler is properly equipped without adequate radar, communication equipment and navigation aids.

HIGH SPEED CRAFT

These may have civil or military application so they are considered before going on to consider specific warship types. A number of hull configurations and propulsion systems are discussed; each designed to overcome problems with other types or to confer some desired advantage. Thus catamarans avoid the loss of stability at high speed suffered by round bilge monohulls. They also provide large deck areas for passenger use or deployment of research or defence equipment. Hydrofoil craft benefit from reduced resistance by lifting the main hull clear of the water. Air cushion vehicles give the possibility of amphibious operation. The effect of waves on performance is minimized in the S mall W aterplane A rea T win H ull (SWATH) concept. Some craft are designed to reduce wash so that they can operate at higher speeds in restricted waters.

The choice of design depends upon the intended service. In some cases a hybrid is used. Although most applications of these concepts have been initially to small craft some are now appearing in the medium size, especially for high speed ferry services. In commercial applications one of the special characteristics, such as those mentioned above, may be the deciding factor in the adoption of a particular hull form. In other cases, particularly for ferries, it may be the extra speed which is a feature of several forms. One way of assessing the relative merits of different forms is what is termed the *transport efficiency factor* which is the ratio of the product of payload and speed to the total installed power.

MONOHULLS

Many high speed small monohulls have had hard chines. Round bilge forms at higher speeds have had stability problems. Hard chine forms with greater beam and reduced length give improved performance in calm water but experience high vertical accelerations in a seaway. Their ride can be improved by using higher deadrise angles leading to a ' deep vee ' form. Current practice favours round bilge for its lower power demands at cruising speed and seakindliness, with the adoption of hard chines for Froude numbers above unity for better stability. One advantage of the round bilge form in seakeeping is that it can be fitted more readily with bilge keels to reduce rolling.

SURFACE EFFECT SHIPS (SESS)

The earliest form of SES was the hovercraft in which the craft was lifted completely clear of the water on an air cushion created by blowing air into a space under the craft and contained by a skirt. For these craft propulsion is by airscrew of jet engines. In some later craft rigid sidewalls remain partially immersed when the craft is raised on its cushion and the skirt is limited to the ends. The sidewalls mean that the craft is not amphibious and cannot negotiate very shallow water. They do, however, improve directional stability and handling characteristics in winds.

They also reduce the leakage of air from the cushion and so reduce the lift power needed and they enable more efficient water propellers or waterjets to be used for propulsion. For naval applications it is usual for the sidewalls to provide sufficient buoyancy so that when at rest with zero cushion pressure the cross structure is still clear of the water. The effect of the air cushion is to reduce the resistance at high speeds, making higher speeds possible for a given power. For ferries, which operate close to their maximum speed for the major part of their passage, it is desirable to operate at high Froude number to get beyond the wavemaking hump in the resistance curve. This, and the wish to reduce the cushion perimeter length for a given plan form area, means that most SESs have a low length to beam ratio. In other applications the craft may be required to operate efficiently over a range of speeds. In this case a somewhat higher length to beam ratio is used to give better fuel consumption rates at the lower cruising speeds. SESs are employed as ferries on a number of short-haul routes.

Early SESs were relatively high cost, noisy craft requiring a lot of maintenance, particularly of the skirts which quickly became worn. As a result of experience with the early, mainly military, craft later versions are considerably improved in all these respects. Naval applications include landing small numbers of covert forces and in mine hunting. In the former, an amphibious craft can cross the exposed beach quickly. The latter use arises from the relative immunity of SESs to underwater explosions.

HYDROFOIL CRAFT

Hydrofoil craft make use of hydrodynamic lift generated by hydrofoils attached to the bottom of the craft. When the craft moves through the water a lift force is generated to counteract the craft's weight, the hull is raised clear of the water and the resistance is reduced. High speeds are possible without using unduly large powers. Once the hull is clear of the water and not, therefore, contributing buoyancy, the lift required of the foils is effectively constant. As speed increases either the submerged area of foil will reduce or their angle of incidence must be reduced. This leads to two types of foil system:

- Completely submerged, incidence controlled. The foils remain completely submerged, reducing the risk of cavitation, and lift is varied by controlling the angle of attack of the foils to the water. This is an ' active ' system and can be used to control the way the craft responds to oncoming waves.

- Fixed surface piercing foils. The foils may be arranged as a ladder either side of the hull or as a large curved foil passing under the hull. As speed increases the craft rises so reducing the area of foil creating lift. This is a ' passive ' system.

Foils are provided forward and aft, the balance of area being such as to provide the desired ride characteristics. The net lift must be in line with the centre of gravity of the craft. Like the SES, the hydrofoil has been used for service on relatively short-haul journeys. Both types of craft have stability characteristics which are peculiarly their own.

MULTI-HULLED VESSELS

These include sailing catamarans, trimarans, offshore rigs, diving support vessels and ferries. Catamarans are not new as two twin hulled paddle steamers of about 90 m length were built in the 1870s for cross channel service. They were liked by passengers for their seakeeping qualities but were overtaken fairly soon by other developments. The upper decks of catamarans provide large areas for passenger facilities in ferries or for helicopter operations. Their greater wetted hull surface area leads to increased frictional resistance but the relatively slender hulls can have reduced resistance at higher speeds, sometimes assisted by interference effects between the two hulls. A hull separation of about 1.25 times the beam of each hull is reasonable in a catamaran. Manoeuvrability is good. High transverse stability and relatively short length mean that seakeeping is not always good. This has been improved in the wave piercing catamarans developed to reduce pitching, and in SWATH designs where the waterplane area is very much reduced and a large part of the displaced water volume is well below the waterline. The longitudinal motions can be reduced by using fins or stabilizers. As a development of twin hull vessels the trimaran form has been proposed. Many design studies indicated many advantages with no significant disadvantages. To prove the concept, and particularly to prove the viability of the structure, a 98 m, 20 knot, demonstrator – *RV Triton* – was completed in 2000. Its structure was designed in accordance with the High Speed and Light Craft Rules of DNV. The main hull is of round bilge form.

The side hulls are of multi-chine design on the outboard face with a plane inboard face. The main hull structure is conventional and integrated with a box girder like cross deck from which the side hulls extend. Propulsion is diesel electric with a single propeller, and rudder, behind the main hull with small side hull thrusters. The trials were extensive and in most cases successfully vindicated the theories. The pentamaran forms are developments of the trimaran with a slender main hull and two small hulls each side. Comparisons of monohulls with multi-hull craft are difficult. Strictly designs of each type should be optimized to meet the stated requirements. Only then can their relative merits and demerits be established. For simpler presentations it is important to establish the basis of comparison be it equal length, displacement, or carrying capacity. Multi-hull designs have a relatively high structural weight and often use aluminium to preserve payload. Wave impact on the cross structure must be avoided or minimized so high freeboard is needed together with careful shaping of the undersides. Because of their small waterplane areas, SWATH ships are very sensitive to changes in load and its distribution so weight control is vital.

RIGID INFLATABLE BOATS (RIBS)

Inflatable boats have been in use for many years and, with a small payload, can achieve high speed. The first rigid inflatables came into being in the 1960s with an inflatable tube surrounding a wooden hull. Much research has gone into developing very strong and durable fabrics for the tubes to enable them to withstand the harsh treatment these craft get. Later craft have used reinforced plastic and aluminium hulls. RIBs come in a wide range of sizes and types.

Some are open, some have enclosed wheelhouse structures; some have outboard motors, others have inboard engines coupled to propellers or waterjets. Lengths range from about 4–16 m and speeds can be as high as 80 knots. Uses are also wide in scope, ranging from leisure through commercial to rescue and military. Users include the military, coastguards, customs and excise, the RNLI, oil companies and emergency services. Taking the RNLI use as an example, the rigid lower hull is shaped to make the craft more seakindly and the inflatable collar safeguards against sinking by swamping.

COMPARISON OF HIGH SPEED TYPES

All the types discussed in this Section have advantages and disadvantages. As stated above for the multihulls, a proper comparison requires design studies to be created of each prospective type, to meet the requirement. However, some special requirement, such as the need to operate over land and sea, may suggest one particular form.

For instance, a craft capable of running up on to a hard surface points to an air cushion vehicle. Many of these types of craft in use today are passenger carrying. SESs with speeds of over 40 knots are common, and can compete with air transport on some routes. Hydrofoils enjoy considerable popularity for passenger carrying on short ferry routes because of their shorter transit times. Examples are the surface piercing Rodriguez designs and the Boeing Jetfoil with its fully submerged foil system. Catamarans are used for rather larger high speed passenger ferries.

YACHTS

For many years, the design, construction and sailing of yachts has been a fascinating art about which whole books are regularly published. This is because the science is too complex for precise solution – and indeed, few yachtsmen would wish it otherwise. Some tenable theories have, in fact, been evolved to help in explaining certain of the performance characteristics of sailing boats, for example, see Claughton *et al.* and Larsson and Eliasson. A yacht, of course, obeys the fundamental theory described generally in this book for all surface ships. In addition, a yacht is subject to air forces acting on the sails and to water forces due to its peculiar underwater shape – forces which are negligible for ordinary surface ships. Sailing before the wind, a yacht is propelled by:

- the vector change of momentum of the wind, deflected by the sails;
- lift generated by the sails acting as aerofoils; because an aerofoil requires an angle of incidence, yachtsmen prefer sailing with wind on the quarter rather than dead astern, particularly when flying a spinnaker, to give them more thrust.

When sailing into the wind, the yacht is propelled only by a component of the lift due to the sails, acting as aerofoils. Lift and associated drag depend upon the set of the sails, their sizes, shapes, stretch and material, the angle of heel of the boat, the relative wind velocity and the presentation of the sails to the wind and to each other. Because the yacht does not quite point in the direction in which it is going and is also heeled, the hull too acts as an aerofoil, experiencing hydrodynamic lift and drag which exactly balance out the air forces when the boat is in steady motion.

The transverse couple produced by air and water forces is reacted to by the hydrostatic righting moment to keep the boat in stable equilibrium. Large angle stability and dynamical stability are clearly of great importance. Longitudinally, the relative position of hydrodynamic lift, the centre of lateral resistance and the lateral component of air lift determine whether the rudder carries weather helm. Lee helm is dangerous because, if the tiller is dropped or the rudder goes free by accident, the boat will not come up into the wind but veer away and increase heel. Ideally, to minimize rudder drag, a yacht should carry slight weather helm in all attitudes and it is towards this ' balance ' that yacht designers aim.

The resistance of a yacht calculated or measured in the manner conventional for surface ships, is not a very helpful guide to the yacht designer. Minimum resistance is required at small angles of yaw and small angles of heel, and these are different from the conventional figures. Resistance in waves is also of considerable importance and varies, of course, with the response of the boat to a particular sea – because of augment of resistance due to pitching, a yacht may well sail faster in Force 4 conditions than it does in Force 5, or better in the Portland Reaches than off Rhode Island. A yacht designer must therefore achieve minimum resistance yawed, heeled and in waves, good longitudinal balance in all conditions and satisfactory stability. The rig must not upset the longitudinal balance and must give maximum performance for sail area permitted, in all conditions and direction of wind, particularly close to the wind. While some theory helps, the process overall remains much an art.

WARSHIPS

Some very interesting problems attach to the design of warships. A fighting ship needs to carry sensors to detect an enemy and weapons to defend itself and attack others. It must be difficult for an enemy to detect and be able to take punishment as well as inflict it. Its ability to survive depends upon its *susceptibility* to being hit and its *vulnerability* to the effects of a striking weapon. Susceptibility depends upon its ability to avoid detection and then, failing that, to prevent the enemy weapon hitting.

STEALTH

A warship can betray its presence by a variety of *signatures.* All must be as low as possible to avoid detection by an enemy, to make it more difficult for enemy weapons to home in and to prevent the triggering of sensitive mines. The signatures include:

- *Noise* from the propulsor, machinery or the flow of water past the ship. An attacking ship can detect noise by passive sonars without betraying its own presence. Noise levels can be reduced by special propulsor design, by fitting anti-noise mountings to noisy machines and by applying special coatings to the hull. Creating a very smooth hull reduces the risk of turbulence in the water.
- *Radar cross section.* When illuminated by a radar a ship causes a return pulse depending upon its size and geometry. By arranging the structural shape so that the returning pulses are scattered over a wide arc the signal picked up by the searching ship is much weaker. Radar absorbent materials can be applied to the outer skin to absorb much of the incident signal.
- *Infrared* emissions from areas of heat. The reader will be aware that instruments are used by rescue services to detect the heat from human bodies buried in debris. The principle is the same. The ship is bound to be warmer than its surroundings but the main heat concentrations can be reduced. The funnel exhaust can be cooled and can be pointed in a direction from which the enemy is less likely to approach.
- *Magnetic.* Many mines are triggered by the changes in the local magnetic field caused by the passage of a ship. All steel ships have a degree of in-built magnetism which can be countered by creating opposing fields with special coils carrying electrical current. This treatment is known as *degaussing.* In addition the ship distorts the

earth's magnetic field. This effect can be reduced in the same way but the ship needs to detect the strength and direction of the earth's field in order to know what correction to apply.

- *Pressure.* The ship causes a change in the pressure field as it moves through the water and mines can respond to this. The effect can be reduced by the ship going slowly and this is the usual defensive measure adopted.

It is impossible to remove the signatures completely. Indeed there is no need in some cases as, for instance, the sea has a background noise level which helps to ' hide ' the ship. The aim of the designer is to reduce the signatures to levels where the enemy must develop more sophisticated sensors and weapons, must take greater risk of being detected in order to detect, and to make it easier to seduce weapons aimed at the ship by means of countermeasures. Enemy radars can be jammed but acts such as this can themselves betray the presence of a ship. Passive protection methods are to be preferred.

SENSORS

Sensors require careful siting to give them a good field of view and to prevent the ship's signatures or motions degrading their performance. Thus search radars must have a complete 360° coverage and are placed high in the ship. Hull mounted sonars are usually fitted below the keel forward where they are remote from major noise sources and where the boundary layer is still relatively thin. Some ships carry sonars that can be towed astern to isolate them from ship noises and to enable them to operate at a depth from which they are more likely to detect a submarine. Weapon control radars need to be able to match the arcs of fire of the weapons they are associated with. Increasingly this means 360° as many missiles are launched vertically at first and then turn in the direction of the enemy. Often more than one sensor is fitted. Sensors must be sited so as not to interfere with, or be affected by, the ship's weapons or communications.

OWN SHIP WEAPONS

Even a ship's own weapons can present problems for it, apart from the usual ones of weight, space and supplies. They require the best possible arcs and these must allow for the missile trajectory. Missiles create an efflux which can harm protective coatings on structure as well as more sensitive equipment on exposed decks. The weapons carry a lot of explosive material and precautions are needed to reduce the risk of premature detonation. Magazines are protected as much as possible from penetration by enemy light weapons and special firefighting systems are fitted and venting arrangements provided to prevent high pressure build up in the magazine in the event of a detonation. Magazine safety is covered by special regulations and trials.

ENEMY WEAPONS

Most warships adopt a policy of layered defence. The aim is to detect an enemy, and the incoming weapon, at the greatest possible range and engage it with a long range defence system. This may be a hard kill system, to take out the enemy vehicle or weapon, or one which causes the incoming weapon to become confused and unable to press home its attack. If the weapon is not detected in time, or penetrates the first line of defence, a medium range system is used and then a short range one. Where an aircraft carrier is present in the task force, its aircraft would usually provide the first line of defence. It is in the later stages that decoys may be released.

The incoming weapon's homing system locks on to the decoy and is diverted from the real target although the resulting explosion may still be uncomfortably close. The shortest range systems are the *close in weapon systems.* These essentially are extremely rapid firing guns which put up a veritable curtain of steel in the path of the incoming weapon. At these very short ranges even a damaged weapon may still impact the ship and cause considerable damage.

SUSTAINING DAMAGE

Even very good defence systems can be defeated, if only by becoming saturated. The ship, then, must be able to withstand at least some measure of damage before it is put out of action completely and even more before it is sunk.

The effects on the ship will generally involve a combination of structural damage, fire, flooding, blast, shock and fragment damage. The ship must be designed to contain these effects within as small a space as possible by *zoning,* separating out vital functions so that not all capability is lost as a result of one hit, providing extra equipments (redundancy) and protection of vital spaces. This latter may be by providing splinter proof plating or by siting well below the waterline. An underwater explosion is perhaps the most serious threat to a ship. This can cause whipping and shock.

VULNERABILITY STUDIES

General ship vulnerability was discussed earlier. Whilst important for all ships it is especially significant for warships because they can expect to receive damage in action with the enemy. Each new design is the subject of a vulnerability assessment to highlight any weak elements. The designer considers the probability of each of the various methods of attack an enemy might deploy, their chances of success and the likely effect upon the ship. The likelihood of retaining various degrees of *fighting capability,* and finally of simply surviving, is calculated. A fighting capability would be a function such as being able to destroy an incoming enemy missile. The contribution of each element of the ship and its systems to each fighting capability is noted, usually in diagrammatic form. For instance, to destroy a missile would require some detection and classification radar, a launcher and weapon, as well as electrics and chilled water services and a command system. Some elements will contribute to more than one capability. For each form of attack the probability of the individual elements being rendered non-operative is assessed using a blend of calculation, modelling and full scale data. If one element is particularly liable to be damaged, or especially important, it can be given extra protection or it can be duplicated to reduce the overall vulnerability. This modelling is similar to that used for reliability assessments. The assessments for each form of attack can be combined, allowing for the probability of each, to give an overall vulnerability for the design. The computations can become quite lengthy. Some judgements are very difficult to make and the results must be interpreted with care. For instance, reduced general services such as electricity may be adequate to support some but not all fighting capabilities.

What then happens, in a particular battle, will depend upon which capabilities the command needs to deploy at that moment. For this reason the vulnerability results are set in the context of various engagement scenarios. In many cases, the full consequences of an attack will depend upon the actions taken by the crew in damage limitation. For instance, how effectively they deal with fire and how rapidly they close doors and valves to limit flooding. Recourse must be made to exercise data and statistical allowances made for human performance. Whilst such analyses may be difficult they can highlight design weaknesses early in the design process when they can be corrected at little cost.

TYPES OF WARSHIP

Warships are categorized by their function, for instance the:

- aircraft carrier;
- guided missile cruiser;
- destroyer;
- frigate;
- mine countermeasure vessel;
- submarine;
- support ship;
- amphibious operations ship.

It is not possible to describe all these types but the following notes will provide a flavour of what is involved in warship design. It will be realized that various ship types usually operate as a group or task force. This will be reflected in their attributes. Thus a frigate will often act as an escort and must be sufficiently fast and manoeuvrable to hold and change station as the group changes course. The task force will exercise defence in depth – a layered defence. If a carrier is present its aircraft will provide a long range surveillance and attack capability. Then there will be long and medium range missile systems on the cruisers and destroyers, decoy systems to seduce the incoming weapon and, finally, a close-in weapon system. The range of sensors provided must match the requirements of the weapons fitted.

Frigates and destroyers

These tend to be maids of all work but with a main function which is usually anti-submarine or antiaircraft. Their weapon and sensor fits and other characteristics reflect this. Usually the main armament is some form of missile system designed to engage the enemy at some distance from the ship, be it an aircraft or submarine. The missile having been fired from a silo or specialist launcher, may be guided all the way to the target by sensors in the ship or may be self-directing and homing. In the latter case, having been directed in the general direction of the target, the weapon's own sensors acquire the target and control the final stages of attack, leaving the ship free to engage other targets. The use of helicopters greatly extends the area of ocean over which the ship can exert an influence.

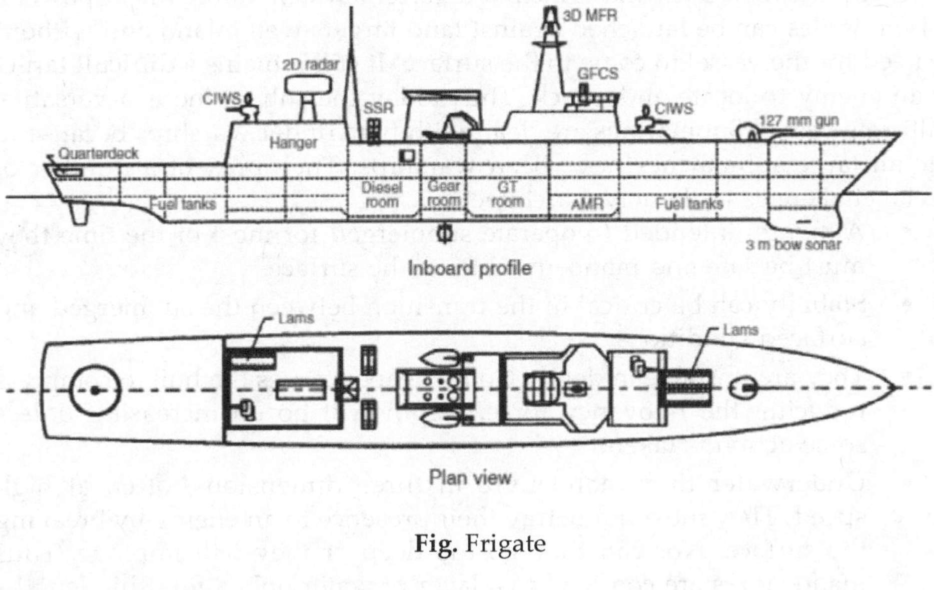

Fig. Frigate

Mine countermeasures vessels

Mine countermeasure ships may be either sweepers or hunters of mines, or combine the two functions in one hull. Modern mines can lie on the bottom and only become active when they sense a target with quite specific signature characteristics. They may then explode under the target or release a homing weapon. They may only react after a selected number of ships have passed nearby, or only at selected times. All these features make them difficult to render harmless. Since sweeping mines depends upon either cutting their securing wires or setting them off by simulated signatures, to which they will react, the latest mines are virtually unsweepable. They need to be hunted, detection being usually by a high-resolution sonar.

They can then be destroyed by placing a small charge alongside the mine to trigger it. The charge is usually laid by a remotely operated underwater vehicle. Because mine countermeasure vessels themselves are a target for the mines they are trying to destroy, the ship signatures must be extremely low and the hulls very robust.

Submarines

The submarine with its torpedoes proved a very potent weapon during the two world wars in the first half of the 20th Century in spite of its limited range of military missions. It could fire torpedoes, lay mines and land covert groups on an enemy coast. Since then its fighting capabilities have been extended greatly by fitting long range missile systems which can be deployed against land, sea or air targets.

The intercontinental ballistic missile, with a nuclear warhead, enabled the submarine to become the principal deterrent system of the major powers; cruise missiles can be launched against land targets well inland and without the need for the vessel to come to the surface. It still remains a difficult target for an enemy to locate and attack. Thus today the submarine is a versatile, multi-role vessel. Submarines are dealt with here under warships because to date all large submarines have been warships. They present a number of special challenges to the naval architect:

- Although intended to operate submerged for most of the time they must be safe and manoeuvrable on the surface.
- Stability can be critical in the transition between the submerged and surfaced conditions.
- They are unstable in depth. Going deeper causes the hull to compress reducing the buoyancy force. Depth will go on increasing unless some action is taken.
- Underwater they manoeuvre in three dimensions, often at high speed. They must not betray their presence to an enemy by breaking the surface. Nor can they go too deep or they will implode. Thus manoeuvres are confined to a layer of water only a few ship lengths in depth.
- Machinery must be able to operate independently of the earth's atmosphere.
- The internal atmosphere must be kept fit for the crew to breathe and free of offensive odours.

- Periscopes are provided for the command to see the outside world and other surface piercing masts can carry radar and communications.
- Escape and rescue arrangements must be provided to assist in saving the crew from a stricken submarine.

Large tanks, mainly external to the pressure hull, are needed which can be flooded to cause the ship to submerge or blown, using compressed air, for surfacing. Propulsion systems are needed for both the surfaced and submerged conditions. For nuclear submarines, or those fitted with some other form of *air independent propulsion* (AIP), the same system can be used. ' Conventional ' submarines use diesels for surface operations and electric drive, powered by batteries, when submerged. An air intake pipe or ' snort' mast can be fitted to enable air to be drawn into the boat at periscope. Batteries are being constantly improved to provide greater endurance underwater and much effort has been devoted to developing AIP systems to provide some of the benefits of nuclear propulsion without the great expense. Closed-cycle diesel engines, fuel cells and Stirling engines are possibilities. The systems still require a source of oxygen such as high-test peroxide or liquid oxygen. Fuel sources for fuel cell application include sulphur free diesel fuel, methanol and hydrogen.

Nuclear propulsion is expensive and brings with it problems of disposing of spent reactor fuel. For these reasons increasing interest is being taken in fuel cells. Having given a submarine a propulsion capability for long periods submerged, it is necessary to make provision for better control of the atmosphere for the crew. The internal atmosphere can contain many pollutants, some becoming important because they can build up to dangerous levels over a long time. A much more comprehensive system of atmosphere monitoring and control is needed than that fitted in earlier conventional submarines. Clancy describes in some detail *USS Miami* (SSN-755) and *HMS Triumph* (S-93) plus the ordering and build procedures, roles and missions. The reference includes the weapons and sensors fitted, dimensions, diving depths and speeds.

Diagrammatic layouts of these submarines (and sketches of others) are given. Anechoic tiling and radar absorbent materials, to improve stealth are mentioned. The hydroplanes, fitted for changing and maintaining depth, can only exert limited lift, particularly when the submarine is moving slowly, so the vessel must be close to neutral buoyancy when submerged and the longitudinal centres of buoyancy and weight must be in line. It follows that the weight distribution before diving, and the admission of ballast water when diving, must be carefully controlled. The first task when submerged is to ' catch a trim ', that is adjust the weights by the small amounts needed to achieve the balance of weight and buoyancy. Since there is no waterplane, when submerged the metacentre and centre of buoyancy will be coincident and BG will be the same for transverse and longitudinal stability. On the surface the usual stability principles apply but the waterplane area is relatively small. The stability when in transition from the submerged to the surfaced state may be critical and needs to be studied in its own right. The usual principles apply to the powering of submarines except that for deep operations there will be no wavemaking resistance.

This is offset to a degree by the greater frictional resistance due to the greater wetted hull surface. The pressure hull, with its transverse bulkheads, must be able to withstand the crushing pressures at deep diving depth. Design calculations usually assume axial symmetry of structure and loads. This idealization enables approximate and analytical solutions to be applied with some accuracy.

Subsequently detailed analyses can be made of non axi-symmetric features such as openings and internal structure. The dome ends at either end of the pressure hull are important features subject usually to finite element analysis and model testing. Buckling of the hull is possible but to be avoided. Assessments are made of *inter-frame collapse* (collapse of the short cylinder of plating between frames under radial compression); inter-bulkhead collapse (collapse of the pressure hull plating with the frames between bulkheads) and frame tripping. The design is developed so that any buckling is likely to be in the inter-frame mode and by keeping the risk of collapse at 1.5 times the maximum working pressure acceptably small.

The effects of shape imperfections and residual stresses are allowed for empirically. Small departures from circularity can lead to a marked loss of strength and the pressure causing yield at 0.25% shape imperfection on radius can be as little as half that required for perfect circularity. If a stricken submarine is lying on the seabed the crew would await rescue if possible. For rescue at least one hatch is designed to enable a rescue submersible to mate with the submarine. The crew can then be transferred to the surface in small groups without getting wet or being subject to undue pressure. The first such rescue craft, apart from some early diving bells, were the two Deep Submergence Rescue Vessels (DSRVs) of the USN. However, deteriorating conditions inside the damaged submarine may mean that the crew cannot await rescue in this way; the pressure may be rising due to water entry or the atmosphere may become polluted.

In such cases the crew can escape from the submarine in depths down to 180 m. Oneand two-man escape towers are fitted to allow rapid compression so limiting the body's uptake of gas which would otherwise lead to the ' bends '. A survival suit is worn to protect against hypothermia and a hood holds a bubble of gas for breathing. In Russian submarines emergency escape capsules are provided. So far commercial applications of submarines have been generally limited to submersibles some of which have been very deep diving.

Many are unmanned, remotely operated vehicles. Most of these applications have been associated with deep ocean research, the exploitation of the ocean's resources, rescuing the crews of stricken submarines or for investigations of shipwrecks. A growing use is in the leisure industry for taking people down to view the colourful sub-surface world. In some types of operation the submersible may be the only way of tackling a problem such as the servicing of an oil wellhead in situ which is too deep for divers.

2

Understand the Regulatory Framework

REGULATORY AUTHORITIES

This chapter sets out the arrangements for enforcement of health and safety in construction activities between the Health and Safety Executive (HSE), the Maritime and Coastguard Agency (MCA) and the Marine Accident Investigation Branch (MAIB). An agreement HSE sets out the roles and responsibilities for each organisation. Under this agreement, wherever there is reference to MCA having a responsibility it shall also be read that MAIB has the responsibility to investigate accidents. Wherever HSE has responsibility, this also includes accident investigation. In principle, HSE has enforcement responsibility for construction work, even carried out on or from a ship. This is without prejudice to MCA.s responsibility under the Merchant Shipping Act for the safety of the ship, the crew and others on board while the ship is used in navigation. When applied to a dredger and a jack-up platform.

Table summarises these in more detail. The legislation, whilst different depending on whether it is HSE or MCA jurisdiction, is converging to risk-based approaches rather than prescriptive requirements. Construction activities covered by the HSE/MAIB/MCA agreement include construction, repair or demolition of marine structures (*e.g.* bridges, piers, jetties, link spans and coastal and flood protection works, piling, dredging, lifting, and also the construction of.non-marine. works which are adjacent to navigable waters, when the work involves support from vessels such as work boats or rescue craft).

Lead responsibility for enforcement and investigations will normally be as follows:

- All aspects of ship safety and operation, including construction and maintenance, manning and emergency arrangements. responsibility of MCA
- Activities associated with pipe-laying, pipeline repair, removal, etc.. responsibility of HSE.

RESPONSIBILITIES OF THE HEALTH AND SAFETY EXECUTIVE

Under HSWA, the HSE is primarily responsible for enforcing legislation, including CDM, covering the occupational health and safety of shore-based workers engaged in a land-based work activity, and work equipment supplied by the shore, for example:

- For workers and equipment (including cranes, etc. which are temporarily mounted on barges) employed on construction work on vessels on inland waters, coastal waters within GB Territorial Waters, where the primary purpose of the vessel is related to land-based works or undertakings. However, the Merchant Shipping and Fishing Vessels (Health and Safety at Work) Regulations 1997 (SI 1997/2962) which applied health and safety legislation to seagoing ships have been amended by the Merchant Shipping and Fishing Vessels (Health and Safety at Work) (Amendment) Regulations 2001 (SI 2001/54). The effect of the amendment is to apply merchant shipping health and safety legislation to non seagoing ships, thereby allowing effective enforcement by the Maritime and Coastguard Agency. However, HSE legislation still covers those workers involved in construction activities.
- For workers and equipment carrying out cargo transfers between ships with one ship moored alongside another, when the cargo transfer involves the use of shore-based equipment and/or staff (including the use of loaders provided by shore-based contractors and cranes which are temporarily mounted on barges).
- For all aspects of health and safety on vessels which are not capable of navigation.

RESPONSIBILITIES OF MARITIME AND COASTGUARD AGENCY

The MCA is an Executive Agency of the Department for Transport and has responsibility and accountability for the UK Merchant Shipping Regulations and their enforcement. It is responsible for the survey and certification of any ships (*i.e.* vessels used in navigation) engaged in coastal and other construction works, their compliance with merchant shipping legislation, the competency and occupational health and safety of the crew (*i.e.* not including workers who are normally shore based). The MCA is responsible for the safety enforcement in the following situations:

- The safety (construction, equipment, navigation) of all vessels (including floating cranes) which are used in navigation.
- The occupational health and safety of anyone on a vessel while it is being used in navigation; while the vessel is being used as an adjunct to construction works, MCA remain responsible for the occupational

health and safety of the crew, but HSE would be responsible for the occupational health and safety of the construction workers.

- For cargo transfers between ships with one ship moored alongside another:
  - in a port or in inland waters when the transfer is carried out solely by the ships. crews under the control of the master(s)
  - at anchor in territorial waters (even if shore-based personnel are involved) since the prime concerns are the safety of the ships and the risk of pollution; also the operations will be carried out under the control of the ships' master(s).

RESPONSIBILITIES OF MARINE ACCIDENT INVESTIGATION BRANCH

The MAIB is responsible for investigating any.navigation. accidents to or on ships engaged in coastal works, including safety vessels and floating cranes (*i.e.* except those investigated by HSE). Its purpose is to investigate accidents at sea so as to determine circumstances and causes with the aim of improving the safety of life at sea and the avoidance of accidents in the future. It is not the purpose to apportion liability, nor, except as far as necessary to achieve the fundamental purpose, to apportion blame.

RESPONSIBILITIES OF PORTS DIVISION OF DFT

Ports Division has a statutory responsibility under the Coast Protection Act 1949 to ensure that coastal works and the siting of offshore installations create no hindrance to navigation. The Act was extended by the Continental Shelf Act 1989 to all parts of the UK continental shelf.

ENFORCEMENT

The MCA and HSE have similar powers of enforcement including, in particular, powers to issue improvement and prohibition notices where HSE inspectors or MCA surveyors observe matters of evident concern in areas for which they are not the lead authority. These should be reported without delay to the appropriate office of the other authority by telephone:

- Where HSE inspectors or MCA surveyors observe matters of evident concern, in areas for which they are not the lead authority, and which pose a risk of serious injury, they may issue a prohibition notice. If they do so, they should notify the appropriate MCA or HSE office by telephone and send a copy of the Notice by fax at the earliest opportunity.
- Where both organisations have an interest, improvement notices should normally be served by the lead authority after consultation with the other authority.

ACCIDENT INVESTIGATION

The HSE inspectors have powers to investigate accidents to shore-based workers. Accidents to seamen or to ships would be investigated by MAIB. The lead responsibility would be allocated on a case-by-case basis on the same criteria as for the division of responsibilities between HSE and MCA. Where misuse of land-based equipment temporarily mounted on a vessel affected its safety, a joint investigation might be necessary.

RELEVANT KEY LEGISLATION

HEALTH AND SAFETY AT WORK ACT (1974)

The Health and Safety at Work Act 1974 (HSWA) covers almost all work situations, and places general duties on employers to provide protection for all people at work and some protection to the general public.

The HSWA is of over-riding importance, and its requirements should be considered before reference to other Acts and regulations that may also be applicable. The HSWA places a number of general duties on employers, including:

- The provision and maintenance of plant and systems of work that are safe
- The provision of information, instruction, training and supervision to ensure the safety of employees at work
- The maintenance of any place of work under the employer's control in a condition that is safe and without risk to health
- The provision of safe and risk-free access to and egress from any place of work under the employer's control
- Ensuring that persons not in the employer's employment are not exposed to risks to their health and safety.

The HSWA also places duties on employees to take reasonable care of their own and others. health and safety. Section 6 of HSWA places duties on designers, manufacturers, suppliers, erectors and installers to ensure that articles are, so far as is reasonably practicable, safe. They must also arrange any consequential tests and examinations, provide adequate information for safe use, and carry out necessary research to minimise risks to health and safety.

The HSWA provides for health and safety regulations to be made under Section 15, and for practical guidance on health and safety to be provided in the form of approved codes of practice under Section 16. Failure to observe the provisions of approved codes of practice without following effective and suitable alternative procedures can be taken as evidence of contravention in criminal prosecutions. This obligation is absolute.

MERCHANT SHIPPING ACT (1995)

The Merchant Shipping Act 1995 is an Act to consolidate the Merchant Shipping Acts 1894 to 1994 and other enactments relating to merchant shipping. It is in 13 Parts and covers the registration and operation of British ships and seamen. Part III. Masters and Seamen, contains a section on safety, health and welfare in terms of conditions for crew. Part IV. Safety, contains general guidance on safety and health on ships. The full text of the Act can be found on the HMSO web site at www.hmso.gov.uk/acts. The Merchant Shipping and Maritime Security Act (1997) contains miscellaneous amendments of the Merchant Shipping Act (1995) as well as extension of powers to deal with emergencies at sea, pollution control and marine safety and maritime security. The full text of the Act can also be found on the HMSO web site.

DOCKS REGULATIONS (1988)

The Docks Regulations 1988 apply to all operations carried out in ports within Great Britain. The essentially identical Docks Regulations (Northern Ireland) 1989 apply in Northern Ireland. Approved codes of practice for each version of the Regulations entitled *Safety in docks* were published at the same time. The codes of practice include useful appendices giving details of other relevant legislation, HSE guidance publications and relevant British Standards, although some of the legislation, publications and standards referred to have been superseded by later documents.

It is important to ensure that the latest versions of the relevant publications and standards are used. The Docks Regulations place duties on employers and employees, including vehicle drivers and equipment operators. Duties are also placed on owners of plant, but not on the master or crew of a ship when working on board without shore assistance, who are covered by a variety of merchant shipping legislation. However, operations involving ship.s crew and dock workers working together are to comply with the Docks Regulations. Operations must be planned and executed to ensure, so far as is reasonably practicable, that no person is exposed to danger. The term.reasonably practicable. requires risk assessment to be carried out and the level of risk to be weighed against the effort needed to introduce remedial measures. The term.practicable. means measures that are possible in the light of current knowledge and invention.

Where more than one person controls operations, as in Ro-Ro operations involving ship.s crew, stevedores and vehicle drivers, collaboration is required in setting up and agreeing strict procedures and systems of work. A reporting system is required for defects, so that timely repairs are carried out; the system of work should be reviewed and modified in the light of any defects. The code of practice requires that appropriate arrangements are made for dealing with emergencies and other unexpected situations, and that appropriate instructions on these arrangements are given in advance. Although the regulations do not apply specifically to ship.s equipment, all dock operations are covered. Those making use of ship.s equipment should plan and execute work safely, and the safety of the equipment should be checked, including the provision of certificates of test or examination. For further advice of ship-to-shore linkspans and walkways refer to CIRIA (1999) (C518). The Dock Regulation 13 requires lifting plant to be used only if it is of good design and construction, of adequate strength for the purpose required, of sound material and free from patent defect, properly installed or assembled, and properly maintained. It is implied that the lifting plant is provided in accordance with a relevant national or international standard and that it is subject to systematic preventive maintenance.

Lifting plant should be subject to regular routine inspection by someone competent to assess its safety. The inspection should cover the operation of limit switches and overload indicators. In addition, lifting appliances should not be used in a manner likely to subject them to overturning, and lifting operations should be stopped when the wind exceeds the specified operating limit, and any additional braking system should then be brought into operation. A calibrated anemometer may need to be installed and an indicator or alarm fitted to warn the operator of excessive wind speeds. Rail-mounted walkways should provide the operator with a clear view of the area adjacent to the wheels, and the wheels should be fitted with guards.

The information provided in the approved code of practice is no longer up to date. New sets of regulations have been introduced that modify the Docks Regulations, and it is the responsibility of the user of the code of practice to ensure that the appropriate regulations and the relevant British or European Standards are being used. It is understood that a substantial revision of the approved code of practice and guidance *Safety in docks* is intended.

MANAGEMENT OF HEALTH AND SAFETY AT WORK REGULATIONS (1999)

The Management of Health and Safety at Work Regulations 1999 (MHSWR) cover general matters of health and safety, and operate alongside many other regulations. Normally, compliance with specific regulations will be sufficient to comply with those elements of MHSWR with which the specific regulations overlap. The regulations require every employer to assess the risks to employees and others who may be affected by the conduct of the undertaking, such as passengers or vehicle drivers embarking on or disembarking from a ship. A written record must be kept of the significant findings of the suitable and sufficient assessment of risks to health and safety. The significant findings should include hazards posing a serious risk, control measures and the extent to which they control the risk, and groups affected by significant risks or hazards. The assessment must be reviewed where there has been a significant change. Appropriate written procedures to be followed in the event of serious and imminent danger, and for danger areas, must be established. These procedures must be implemented by sufficiently competent persons. The procedures should include the requirements of specific regulations. Competent persons and employees must have sufficient training and experience or knowledge of maritime and coastal engineering to implement the procedures.

Employees must be provided with information on the risks to health and safety, the preventive and protective measures, the emergency procedures, and the identity of competent persons. A key feature of these regulations and other Health and Safety legislation is the requirement for competent persons to be appointed. It is important to note that competency within construction of land-based works may not mean that they are competent to work on coastal and maritime projects.

CONSTRUCTION (DESIGN AND MANAGEMENT) REGULATIONS (1994)

The Construction (Design and Management) Regulations 1994, generally known as the CDM Regulations, came into force on 31 March 1995. An approved code of practice was also published in 1995 and updated in 2002. The regulations apply to *construction projects* which employ more than four workers, or last longer than 30 working days or 500 person-days. This includes any such work carried out in GB Territorial Waters. The Construction (Health, Safety and Welfare) Regulations 1996 (CHSWR) apply to all construction work including that within GB Territorial Waters. Construction clients may appoint one or more competent persons to assist them to fulfil the requirements of the regulations.

A competent person can be either an employee or a specialist organisation. Such persons should be given sufficient time and resources to carry out their duties. The appointment of these assistants does not absolve construction clients of their responsibilities under HSWA. The regulations apply to construction work, including building and civil engineering work and alteration, cleaning, repair, maintenance, demolition or dismantling of a structure; site clearance, site investigation, excavation and laying foundations for a structure.

They also apply to the installation, commissioning, cleaning, maintenance and repair of mechanical, electrical, hydraulic or similar services normally fixed within or to a structure including ship-to-shore linkspans and walkways. A structure is any building, steel or reinforced concrete structure, dock, bridge or other similar structure; any formwork or scaffolding; and any fixed plant involving a risk of falling more than 2 m, but only in respect of installation, commissioning, maintenance and decommissioning and dismantling. The client has specific duties under the CDM Regulations, one of which is to appoint a planning supervisor and a principal contractor. The client must be reasonably satisfied with the competence of these appointees and with the adequacy of resources allocated by the appointees to perform their duties. The client must provide health and safety information about the project, not permit construction work to start until an adequate construction health and safety plan is available, and hold the health and safety file and make it available to others.

The client must also issue a notification of the project using Form F10. The client is responsible for providing information about the site. In the case of marine structures, ship-to-shore linkspans and walkways this might include information on soil conditions, seabed levels, wave climate, ship operations and geometrical requirements. On appointment, one of the designer.s duties under the regulations is to advise clients of their duties. Titles that provide guidance on the CDM Regulations are listed in the References, in Chapter and Appendices 2 and 3.

Particularly relevant guidance is provided in HSE information sheet 39, Construction (Design and Management) Regulations 1994. For the role of the client, view the HSE web site at www.hse.gov.uk/pubns. A key feature of the CDM regulations and other Health and Safety legislation is the requirement for competent persons to be appointed. It is important to note that competency within design, construction, cleaning or demolition of land-based works may not mean that they are competent to work on coastal and maritime projects. The CDM Regulations require the designer to weigh up many factors as they prepare their design whilst giving due consideration to health and safety. Only a competent designer, who understands the particular construction processes associated in coastal construction, will be able to address health and safety issues appropriately.

PROVISION AND USE OF WORK EQUIPMENT REGULATIONS (1998)

The Provision and Use of Work Equipment Regulations 1998 (PUWER) implemented EU Directive 89/655/EEC (UWED) and covers health and safety requirements for the provision of safe work equipment and its safe use. The regulations came into effect in 1993, except for equipment that existed before this date, which were covered by a variety of old regulations. For existing equipment, the regulations came into effect in 1997. Existing work equipment sold second-hand since 1993 becomes.new. equipment for the purposes of the regulations, and the requirements of Regulations 11 to 24 apply retrospectively.

Guidance on the regulations was published by HSE in 1992 in *Work equipment. guidance on regulations*. Many of the requirements of PUWER were already covered by earlier legislation and regulations, and several parallel regulations were enacted in 1992, such as MHSWR and the Workplace (Health, Safety and Welfare) Regulations. Where the various regulations set out duties that overlap, compliance with the more specific regulation will normally be sufficient to comply with the general requirement in other regulations. The EU Directive 95/63/EC, the Amending Directive to the Use of Work Equipment Directive (AUWED), lays down wide-ranging requirements for the provision, management and use of mobile, self-propelled and remote-controlled work equipment, inspection of work equipment, provision and use of lifting equipment, and the management of lifting operations. The provisions of PUWER 98 operate alongside the Dock Regulations 1988.

LIFTING AND PULLING LEGISLATION

Lifting Operations and Lifting Equipment Regulations (1998) and Provision and Use of Work Equipment Regulations 1998 (PUWER 98) The LOLER regulations cover.lifting equipment.. The PUWER regulations cover winches used for.pulling.. High-capacity winches are often employed in the pull out of pipelines and outfalls, positioning barges in offshore locations.

They do not cover the vessel.s own lifting equipment, which are not used for construction activities, as these come under different legislation. The Lifting Operations and Lifting Equipment Regulations 1998 (LOLER) implement the specific lifting aspects of AUWED, and apply over and above the general requirements of PUWER 98 to cover specific risks arising from lifting loads. There were numerous existing regulations covering lifting equipment, of which the most relevant to linkspans and walkways were the Docks Regulations 1988 (Regulations 13 to 17) and the Lifting Plant and Equipment (Records of Test and Examination) Regulations 1992.

The lifting equipment aspects of all these regulations are amended or revoked by LOLER. Regulation 14 amends Regulation 13(4) and revokes Regulations 14,15,16(3)(4)(5)(7) and (8) and 17 of the Docks Regulations 1988. One problem of the various existing regulations is that they do not provide a common definition of lifting equipment. The ports industry generally adopts the definitions in the Docks Regulations. The AUWED refers to.work equipment for lifting loads. but gives no definition. Throughout LOLER.lifting equipment. covers cranes and accessories, including slings, hooks and eyebolts. Generally LOLER cover the proper planning of lifting operations so that they can be carried out safely, the safe use of lifting equipment, and the thorough examination of lifting equipment at suitable intervals by a competent person.

The regulations in LOLER are relatively concise, and in order to fully implement AUWED they have been amplified by an approved code of practice giving comprehensive guidance, which also amplifies relevant clauses in PUWER 98. The Management of Health and Safety at Work Regulations already require risk assessments of work activities to be carried out to identify risks arising from the use of lifting equipment and the appropriate precautions required to deal with these risks. The level of precaution will depend on the degree of risk assessed, and should reduce the risk to as low a level as is reasonably practicable. Particular hazards to be considered are equipment striking a person or object and the consequences of equipment failing. The LOLER impose further requirements on employers, on persons controlling the lifting equipment, and on the operator of the lifting equipment or the method of operation of the lifting equipment.

The LOLER define lifting equipment as work equipment for lifting and lowering loads. The enforcement agency has confirmed that LOLER will not be applicable to the majority of linkspans and walkways where the lifting machinery is an integral part of the installation and only adjusts the vertical location of one component of the machine rather than raising or lowering a separately identifiable load. The general exclusion of linkspans and walkways applies both to those that are adjusted to a level then.spragged.

before vehicle or pedestrian loads are applied, and to those that are adjusted in level during use, because the primary function in both cases is not the lifting and lowering of loads. All linkspans and walkways are considered to be work equipment, and are subject to the requirements of PUWER 98. (However, note that where a ship provides a linkspan or walkway it will come under MCA.s jurisdiction.) New guidance on linkspans is given in BS 6349 Part 8. The provisions of LOLER (1998) operate alongside the Docks Regulations 1988.

Regulations for shipping

The PUWER and LOLER refer solely to the regulations introduced by HSE. Similar regulations applicable to shipping have not yet been introduced but are currently out to consultation. In the meantime the following continue to apply in relation to shipping:

- The Merchant Shipping (Guarding of Machinery and Safety of Electrical Equipment) Regulations 1988
- Regulation 10 of the Merchant Shipping (Safe Movement on Board Ship) Regulations 1988
- The Merchant Shipping (Hatches and Lifting Plant) Regulations 1988.

DIVING AT WORK REGULATIONS (1997)

The Diving at Work Regulations 1997 (DWR) and the ACOP for Inland/ Inshore Diving operations apply if the activity occurs:

- Inshore within United Kingdom Territorial Waters adjacent to Great Britain (generally 12 nautical miles from the low water line)
- Inland in Great Britain including docks, harbours, rivers, culverts, canals, lakes, ponds and reservoirs.

Outside this scope, a diving operation is likely to be covered under the Approved Code of Practice (ACOP) covering Commercial Diving Projects Offshore. For the purposes of this publication, all references are based on the assumption that the Inland/Inshore ACOP applies. Regulation 4 states that:

- Every person who to any extent is responsible for, has control over or is engaged in a diving project or whose acts and omissions could adversely affect the health and safety of persons engaged in a such a project, shall take such measures as is reasonable for a person in his position to take to ensure that these regulations are complied with.

In a coastal and maritime environment this could include the employer, a port or harbour master, vessel master, main contractor, designer, specialist service supplier or any operator instructed to use plant or equipment in support of the diving or associated operation. Whilst the level of responsibility will vary from person to person, all should retain an interest in the diving operation and will need to maintain an element of liaison with the diving contractor and their appointed supervisor.

OTHER UK LEGISLATION IN DOCKWORKS

There are also numerous by-laws and local requirements that can be applied by, for example, Port Authorities (ports), Harbour Authorities (coastal works), British Waterways (canals), and Local Authorities (inland lakes), depending on where the work is situated. Work in a water-filled quarry would also be subject to the Mines and Quarries Act. Different responsibilities apply to the different categories of personnel. The HSE provides guidance on managing health and safety in dockwork.

OTHER INTERNATIONAL SHIPPING LEGISLATION/CONVENTIONS

There is extensive legislation that covers safety of shipping with the International Maritime Organisation (IMO) providing overall governance. For instance, there is a system of classifying ships based on various international maritime conventions (*e.g.* SOLAS, MARPOL, etc.). This is carried out by Classification Bodies, like Lloyds Register and Bureau Veritas through audits of the vessels. The ISM (Institute for Safety and Health Management) code is another registration scheme for vessels over 500 tonnes. The overall safety and sea worthiness of vessels of all Flag States within UK waters is the responsibility of the MCA.

ENVIRONMENTAL LEGISLATION

Occupational health is governed both by health and safety legislation listed above and by environmental legislation. Further guidance on environmental legislation is given in CIRIA (2003).

3

Severity, Probability and Risk of Accidents

The review of accidents started with the perception from many stakeholders that coastal and maritime engineering presents a particularly hazardous environment for the construction industry. Simm and Cruickshank identified many of the commercial and technical hazards experienced by key stakeholders within the industry.

However, this earlier review did not focus on health and safety issues. The current review seeks to examine what safety statistics exist and answer the following:

- Is there evidence to suggest that coastal construction has a better or worse accident rate compared to general land construction?
- Is there evidence to suggest that coastal construction has a better or worse accident rate than offshore operations?
- What are the main causes of accidents in coastal and maritime engineering?

Answers to the above can be used to help evaluate cultural, organisational and humans factor competency issues and to identify how these vary between the different sectors. The causes of accidents were evaluated using:

- official data from the HSE RIDDOR database
- data supplied by contractors
- structured interviewing, drawing on expert opinions with respect to the key hazards and the differences between this sector and general and offshore construction.

THE BUSINESS CASE FOR BETTER HEALTH AND SAFETY MANAGEMENT IN COASTAL AND MARITIME ENGINEERING

Work-related accidents in the UK cost industry over £11 billion each year. The coastal and maritime engineering sector constitutes a significant industry and makes therefore a significant contribution to this cost. Apart from the social benefits, avoiding work-related accidents improves the business case with other benefits such as:

- achieving a greater certainty of outcome
- improved planning and control, reduced delays
- more reliable cost and completion dates
- reduced overall costs.

CLIENT FEEDBACK

THE ENVIRONMENT AGENCY

The Environment Agency uses its framework contractors and specialist subcontractors to undertake coastal defence works. In total there are probably only a dozen companies undertaking coastal engineering work for the Environment Agency. Agency staff consulted considered that this helps to maintain high competence levels within the contractors they use. It is the Agency.s stated policy objective to achieve a position of zero reportable construction incidents and to be considered to demonstrate best practice as a client.

The Environment Agency employs both proactive and reactive systems in managing health and safety risk, which is co-ordinated through a national health and safety advisor. Its proactive systems include:

- health and safety pre-qualification processes for prospective contractors (including company procedures and staff competency levels)
- vetting of method statements
- site visits by the EA health and safety advisor
- site meetings (where Health and Safety is a key agenda item)
- quarterly reviews, which include the recording of H&S key performance indicators as one of the measures for assessing contractors. performance and used in the future allocation of work
- post project evaluation (with Health and Safety being a key performance indicator within the evaluation).

On the reactive side, the Environment Agency maintains a database of information on health and safety information relating to its staff, contractors and the public. The purpose of the database is to ensure that accident records (including near misses) are investigated and lessons are learnt thereby improving the Agency.s ability to examine why incidents occur and provide better protection and a safer working environment. Unfortunately, the database had not been operational for sufficient duration to make use of the data. Anecdotal evidence suggests that the Agency.s health and safety record is better than the construction norm. It was suggested that the main reason for this was that there was a general absence of construction work associated with construction at height and the corresponding risks associated with that work. The Agency is also endorsing the Construction Skills Certificate Scheme and will require all contractors. staff to have attained the necessary level of certification.

PORT OPERATOR

The port operator contacted owns and operates two major commercial ports. Its experience identified that the commercial port operations themselves were much more hazardous than the construction operations.

CULTURAL, ORGANISATIONAL AND POLICY FACTORS

A series of structured interviews were held with marine contractors and other stakeholders to draw out cultural, organisational and policy factors that contribute to safety practice in coastal engineering. These structured interviews covered the following principal topics, focusing on the differences between land-based construction and coastal/maritime construction:

- perceived areas of special risk in marine construction
- training/instructions/competence of the workforce
- safety records
- preparation of risk assessments and method statements.

The structured interviews highlighted that:

- Almost all contractors consulted believed that there was generally a higher level of competency on maritime and coastal engineering projects than for general land-based construction. They considered that this high level of competence helped to keep the number of accidents lower and that this may be the reason why accident statistics do not show an increased accident rate for maritime and coastal engineering projects, which are potentially more dangerous.
- In addition, the experienced maritime/coastal engineering contractor will have planned and understood the full scope and risks of the project at the time of tender, and this can generally be recognised in the tender submission, even though this may place the contractor at a disadvantage, if price alone is the driving consideration.

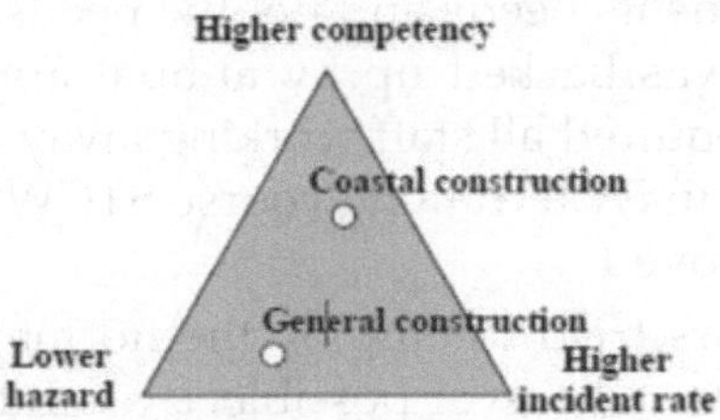

- Safety is often.engineered in. at an early stage in a project because of the close involvement of engineers in the methodology of constructing the works. Typically, in maritime/coastal construction large elements and large forces are in play, be it stability of dock walls or other structures in the water, floating construction plant and other marine craft, temporary works and access structures and others. These demand the attention and skills of the engineer to provide appropriate and safe solutions to the problems of construction in such an environment.

- Almost all of the consultees use direct labour, rather than perhaps the more usual practice in land-based civil engineering of using subcontractors or labour/staff hired for the project. The contractors considered that maritime/coastal work was more specialised typically than that associated with general land-based construction and therefore a higher level of competency was required. The use of direct labour was the only way to realistically achieve this increased competence. Where subcontractors were employed, these tended to be specialised contractors who were well used to working in this environment. The contractors did not accept that the use of management contractors was appropriate in coastal/maritime engineering.
- If the contractors. workload did not entirely consist of coastal or maritime works, then when a coastal engineering project was awarded, the core coastal/maritime engineering team remobilised from the various site teams to work on the project. The core team then trained any new staff to ensure that competency levels remained high throughout the site team. One contractor stated that the average age of their specialist core team was 43, suggesting that experience was a key element of the competency resource.
- Competence of staff was demonstrated using skills training schemes such as the NVQ, Construction Skills Certificate Scheme (CSCS) run by CITB or other apprenticeship schemes. However, many found that standard training schemes did not address their specialised coastal/maritime engineering training needs and that they had to carry out more targeted health and safety training focused towards coastal/maritime engineering themselves. One contractor partnered with one training organisation to ensure that the safety training was targeted towards its needs and not the needs of general construction. This training was backed up by annual appraisals. One contractor stated that it required all staff working away from shore to undertake a one day sea survival training course STCW 95 Approved or OPITO Training Approved.
- Most contractors tried to employ the normal approach of removing the risk at source wherever possible, *e.g.* removing marine plant risks by working from land, removing wave height risk by using a jack-up barge rather than a floating barge where possible, clearly fencing off water edges.

- Consistent with land-based construction, a combination of measures was used to maintain competence such as formal off the job training, on the job training and tool box talks. Also consistent with land-based engineering, method statements and risk assessments were prepared to ensure that the work was undertaken in a planned and safe manner.
- One specialised contractor stated that its management structure was different to a normal land-based construction site and that it had more senior and more highly skilled staff with little to no unskilled staff. It stated that this higher level of competence helped manage safety on its sites.
- One specialised marine contractor stated that it tended to source operators from other similar industries such as the fishing industry who were used to working in maritime environments.
- One specialised marine contractor stated that on its jack-up platform it tended to work in close knit small teams, who each know each other.s roles, responsibilities and competencies. It believed that this working relationship helped manage health and safety.
- A number of contractors stated that apart from the more obvious risks, coastal working often involved working in remote areas with long stretches of coastline and only a few operators spread over significant distances. Therefore, the risk management approach needed to recognise the different working environment such as hazards associated with lone working and remote access.
- A number of contractors commented on the fact that many marine contractors are foreign owned and often operated by foreign personnel resulting in greater variety in health and safety practices. On the positive side the maritime industry benefits from greater cross-fertilisation of ideas but has the disadvantage that cultural differences, differences in legislation and its implementation and communication problems can affect health and safety.

SUMMARY AND CONCLUSIONS

Given the potentially more hazardous environment, it is noticeable that coastal and maritime engineering appears to have similar levels of incident rate to general construction. A general conclusion would be that better organised staff who are more highly trained and have experience in the coastal/maritime sector tend to have fewer accidents. However, all those consulted acknowledged that in addition to the natural hazard issues present in coastal and maritime engineering, many organisational, cultural and other factors influenced health and safety management in coastal and maritime engineering.

4

Risks Issues

This chapter aims to identify and describe key sources of hazards in coastal and maritime engineering. Much of this work is based on Simm and Cruickshank and further information on these issues can be found in that publication. The key sources of hazards derive from:

- The uncertain sea (wind, waves, currents, water levels)
- The physical dynamic environment (impacts due to the above)
- The users of the coastal environment (the lack of containment of the site)
- The selection and use of equipment in these environments
- Impact on the operators (inadequate welfare provisions).

THE UNCERTAIN SEA

The coastal/maritime environment can be highly volatile and at times unpredictable, presenting a major challenge to those engaged in construction work. Conditions can also change very suddenly. This section describes the basic parameters that can influence the planning and execution of construction works in such environments, namely:

- Wind climate
- Wave climate
- Currents
- Water levels
- Joint probabilities of the above.

These basic parameters are discussed briefly below and further details are available from CIRIA, CIRIA/CUR Rock Manual and Simm and Cruickshank. It is important to recognise that waves and tidal action in particular can introduce irregular and varied directional forces.

For example at peninsulas where estuaries meet tidal currents, wave patterns are not predictable and the direction of flow can change at any depth and time, making operations in such areas particularly hazardous. Contractors need to be fully informed on the weather climate at the site. Basic data on prevailing conditions can be obtained from a number of sources. In addition, recent developments have allowed for real-time forecasting of weather conditions, giving contractors three.five day forecasts which can be site specific and allow for forward planning of weather dependent operations such as towage and positioning of caisson units, heavy lifts using floating cranes, etc.

WIND CLIMATE

For marine operations, the wind strength is described in terms of the Beaufort Scale. Apart from the influence that the wind climate can have on the wave climate, winds can also have an important effect on floating craft and on the operation of certain items of contractor plant. The effect of strong and rapidly varying wind speeds and local pressures can cause significant motions of marine craft, which have consequential effects on safety. In addition, wind can have a significant effect on lifting operations, both on shore and, more particularly, on floating crane barges and cranes on jack-up platforms.

Around most of the UK coast, sequential wind records in digital format are available from about 1970 onwards. It should be noted that the coastal environment is much more exposed than many land sites and this has a direct consequence in wind speeds over the water being higher than speeds over land for any given weather condition. Simm and Cruickshank suggested that the increase is of the order of 10%. Further advice on managing issues in the maritime/coastal construction environment is also given in that publication.

WAVE CLIMATE

Waves provide probably the single most important hydraulic parameter in coastal/maritime engineering construction. Wave conditions are described in a number of technical terms. Their impact on construction activities is wide ranging and can:

- Affect deliveries reaching the site (*e.g.* preventing barges leaving port, causing them to run for cover, delaying them in transit, preventing them from unloading)
- Damage plant (due to beaching, overturning, striking the works, etc.)
- Damage temporary works
- Result in personal injuries to operatives through wave impact
- Compromise effective rescue procedures
- Damage completed works where permanent protection is not yet in place
- Draw down beach levels which can affect both the works and deliveries.

The impact of waves is dependent on factors, such as:

- Wave height
- *Wave period:* long-period waves tend to cause more problems for vessel movements and greater overtopping and run up
- Wave length. a function of water depth and wave height
- Likely extreme values of the above factors over a period of one season, one year or more
- Wave direction
- Storm duration
- Storm persistence and sequencing.

Ship-generated waves

In addition to wind-generated waves, consideration must be given to the possible impact of waves generated by passing ships. This can be particularly important in the case of high-speed catamaran ferries, where large bow waves can be generated which may arrive at a site some considerable time (which may be up to 20 minutes) after a vessel has passed.

Wave overtopping

In addition, consideration should be given to wave overtopping hazards. Overtopping can occur on a regular basis as a foreseeable gentle.splash. or on a rarer and often more dangerous and violent basis as a result of.freak. wave(s). Forces that arise from violent overtopping can be well over 200 kN/ m^2. Research is currently underway to examine overtopping so that better understanding and predictions of violent overtopping can be assessed.

TIDAL CURRENTS

Except in the case of construction using caissons or pitching and driving piles in strong currents, the impact of currents on construction risk is generally limited mainly because they are more predictable, especially around the UK coast.

However, in some cases, for example at river or loch entrances or sections of narrow waterways, currents can be high and sometimes unpredictable due to the influence of surges. This is particularly important when considering submerged construction activities using remote operated vehicles or (ROVs) or divers. Currents can:

- Knock personnel over
- Affect the ability of a vessel to hold position offshore
- Affect the ability of a vessel to safely approach the site especially in restricted water depth
- Affect the rescue procedures and ability of rescue craft to operate

- Affect the ability to place materials within tolerance
- Affect the ability to pitch piles and require immediate pile bracing after driving
- Erode partially completed works
- Apply loading on temporary works
- Affect the incident wave conditions.

Currents can be described over a given time-scale by a time.mean magnitude, direction, and spatial and turbulent variability. Tidal currents, like astronomical tidal levels are predictable but can be particularly influenced by less predictable components such as wind, river discharge and surge effects. When wind and wave influences are important in current generation, it is rare for the resulting, albeit less predictable, currents to be a governing construction consideration when compared with the direct wind and wave effects. Waves, water levels and currents can interact to change the conditions of each.

WATER LEVELS

Water levels are important in a number of ways as they may:

- Define what works can be carried out in the dry and the access time available to those works
- Prevent deliveries from reaching the site
- Flood the works where excavations or cofferdams are used
- Affect the wave and current climate that reaches the site
- Affect ground water levels and pore water pressures behind quays and retaining structures.

There are various components of water level which need to be considered. As well as astronomical tides there are several meteorological components of water level. For construction works these residual components comprise:

- Storm surges
- Wind set-up
- Wave set-up
- Seiches.

The major effect determining water level at any one instant, around the UK at least, is the astronomical tide which can be predicted accurately well in advance. Meteorological and seismic effects are not predictable more than, at best, a few days in advance, and even then the predictions are very uncertain. It is these aspects that can pose a risk to the construction works. Changing the operation to work in the dry through the use of limpet dams or caissons can remove the hazards associated with changing water levels. Care should still be taken to ensure inundation does not occur due to excessive water levels, waves, ground water or other means.

JOINT PROBABILITY OF WAVES AND WATER LEVELS

Overtopping and damage to construction works are usually associated with times of both large waves and high water levels. Therefore it can be important to consider both parameters in the design of temporary works. High currents can also occur at these times. Further advice is given in Simm and Cruickshank (1998).

DYNAMIC PHYSICAL ENVIRONMENT

The previous section discussed how the.uncertain sea. affects construction works. This section shows how the presence of the sea interacts with physical and other factors to affect the buildability of coastal works. It illustrates with examples the inherent problems in constructing in the coastal environment and, in particular, discusses the following:

- Ground conditions and changes in seabed and beach profiles
- Uncertainty of condition of existing structures
- Access and working space
- Temporary instability of the works
- Ordnance
- Subsea services
- Timing of the works
- Contaminated sediments
- Interaction of the above.

THE SURF ZONE

Working in the surf zone can be particularly hazardous. The issues to consider include:

- Breaking waves can capsize small vessels especially if power is lost. The vessel can then turn broadside and be at its most vulnerable. For larger vessels, using spuds and anchors to hold position can mitigate this risk. Strong tidal crosscurrents can cause additional problems.
- A manned safety boat in the surf zone is a major hazard and its use should be carefully planned.
- There is limited tug access (*i.e.* restricted to shallow draught tug which means limited power to act in emergencies). Operations often rely on manoeuvring with ropes using winches, two sea moorings and two land moorings.
- The underkeel clearance is less and vessels can ground in a larger swell. There may be additional obstructions resulting from operations (*i.e.* rock placement). Correct vessel selection is important when working in this area.

- Access into intertidal zone (*e.g.* to unload rocks) becomes more difficult when the tidal range is small.
- Weather reports from established sources are useful but in the surf zone waves are all important. Local knowledge is useful and an inshore forecast can be obtained from specialists sources.
- Man overboard procedure and emergency evacuation will be different in the surf zone.

GROUND CONDITIONS AND CHANGES IN SEABED AND BEACH PROFILES

Unlike most civil engineering works, th*e.g*round. surface (seabed and beach) in coastal engineering can be highly mobile, changing from day to day and from season to season. These changes can be caused by:

- long-term trends within the coastal region
- sea-state fluctuations
- summer/winter or storm/swell fluctuations in the position of a coastal bar
- scour/accretion due to presence of plant, temporary works or partly completed permanent works
- wind variations
- dredging adjacent to the shore.

These changes can affect the construction works in a variety of ways as they alter:

- the exposure to wave attack
- the dry time. available at the works and for land access to and from the works
- draught available for marine plant at the works and in the access route to and from the works
- the nature of the surface layers affecting the works and access (*e.g.* sand layer removed to expose soft clay)
- the stability of the permanent works
- the stability and design criteria used in the design of the temporary works.

As with all construction projects, coastal engineering projects will always benefit from good quality site investigations appropriate to the project. Ground condition issues most frequently affect coastal engineering construction. Some are detailed below.

Settlement

Direct settlement of soft ground under loading can be of unacceptable proportions. Ideally, this should be a matter for design. In practice, natural variability of the ground or lack of knowledge of the founding materials may well mean some hazard is left to be determined at the construction stage. Quicksand, particularly on open flat sandbank areas, can be especially hazardous.

Variability in founding levels for structures

The required founding level may be dictated, for example, by the requirement to reach:

- A certain geological stratum (*e.g.* chalk or clay underlying beach material)
- Soil of a particular resistance to loading (*e.g.* resistance to pile-driving loads)
- Soil of a particular density (*e.g.* granular soils of sufficient density, indicated by SPT or CPT values, that will not liquefy under earthquake loading).

Natural variability between boreholes or trial pits, even with the best methods of interpretation (*e.g.* interpolation using seismic traces) will still leave uncertainty here. Indeed, the action of the sea between the design survey and construction can actually affect shallow underlying material. For example, the motion of the beach material under wave action can abrade clay or soft chalk platforms underlying the beach surface.

Obstacles and hard ground

Obstacles and hard ground can affect excavations to required depths, such as in navigational channel dredging projects, or steel piles requiring to be driven to a certain level. Obstacles frequently encountered include old, temporary or abandoned permanent works and sunken vessels. In such cases, obtaining detailed local knowledge or a discussion with the relevant authorities such as the Harbour Master can often reveal much information.

Soft ground

Soft ground may also create problems due to a requirement for flatter back slopes to excavations and by the lack of a firm platform for the operation of plant. Unconsolidated material can result in quicksand developing especially on flat estuarial areas. This is especially the case where dense sand material overlies soft material.

Instability of cliffs

Significant practical problems and safety issues arise when working at the base of cliffs. It should be noted that the risks of dealing with such unforeseen ground conditions are magnified in the marine environment, due to the effects of constant marine action at the base of the cliffs.

Changing soil properties

Soil properties may change from one tide to the next as wave and current loading may liquefy or mobilise seabed and beach material. Therefore, ground which is stable for an excavator one day may alter and become unstable at the next low tide.

Contaminated ground conditions

Contaminated sediments often exist around port and industrial areas. Like land-based construction ground and sediment, contamination can affect human health. These issues are discussed in more detail in CIRIA.

UNCERTAINTY OF CONDITION OF EXISTING STRUCTURES

This can cause three main categories of hazard.

- Hazards similar to those arising from the uncertainty in regard to ground, where the existing structure is being modified in some way as part of the works.
- Hazards arising from failure of.support. where a structure is effectively forming part of the contractor.s temporary works. This frequently occurring situation arises where an existing structure is used to support plant and equipment, and the load which it can safely sustain is difficult to quantify. As in many situations, where such structures make the difference between working in the wet or in the dry, there is great incentive for contractors to use them and accept the hazard (even if unclear about its consequences). The hazard may be defined as that of either collapse, or the structure becoming unusable, under the range of combinations of loads which might be experienced during the course of the contract arising from either the operation of plant, temporary structures or from the weather.
- Hazards of the unknown extent of collision damage due to impacts arising from reduced navigational control of floating plant.

ACCESS AND WORKING SPACE

Available land access routes are normally well defined prior to commencing construction. However, narrow streets leading to the site, a limited number of access points and tidal restrictions may compromise access. In addition, access restrictions may be imposed on a contractor for reasons relating to the public, and public access, recreational and leisure considerations and tourism issues.

Available working space may change with the tide and may be limited especially at high tide. Movement of beach material to create local berms can increase tidal access. Moreover, working areas can be very long and narrow (sometimes several kilometres long), and, like road maintenance projects, can make supervision of health and safety difficult to manage. Marine access may be well defined, but is subject to all the vagaries of the weather affecting draught, navigability, etc. and the risks of seabed debris fouling or damaging craft. Marine access may also be subject to receipt of permissions, the carrying out of surveys prior to and after construction, and concluding appropriate compensation agreements with commercial interests, notably fishing and aquaculture interests. Failure to achieve any of these can lead to significant costs and time delays. Tidal working can also put pressure on operators to finish work before the incoming tide prevents work being completed. This pressure can, if not checked, lead to the disregard of health and safety measures.

TEMPORARY INSTABILITY OF THE WORKS

Unless it is economic to provide cofferdam arrangements (rarely the case with seawalls), excavations may often have to be left open and works left exposed between successive low waters allowing the ingress of the tide at high water. Scour under, or in front of, temporary works/plant can cause major delay and cost impacts. These impacts can include:

- land-based plant becoming temporarily bogged down on a beach. If trapped by the rising tide the damage to the machine can be severe and in the worst case may involve a write-off
- consequential environmental damage through diesel/oil spill from machines which are inundated
- damage due to undermining of vessels temporarily beached on what was originally a flat beach or seabed
- erosion around or accretion against piled cofferdams, potentially causing failure.

The designer generally focuses on the extreme wind, wave, water level and current conditions appropriate to the design of the permanent works but should also take into account the wave conditions that will affect the works during partly completed conditions. The temporary instability of the works during the construction (*e.g.* before primary armour is placed on a breakwater) can result in catastrophic failures and risk to personnel. It is essential that adequate consideration be given to the critical stages during the construction, when the works may be susceptible to damage from relatively mild wind, wave, water level and current conditions.

The temporary instability of the caissons was a fundamental risk for the contractor on one caisson project. Here, the contractor managed this risk by carrying out numerical wave modelling to establish a management system which used UK Meteorological Office offshore wind and wave predictions to give local wave predictions at the site. The site team used the system to assist making decisions as to when to commence, accelerate or terminate operations and to make safety decisions during critical operations. Temporary instability of rock storage areas is another potential hazard. Often these mounds are dumped by marine plant or lorries prior to placing in the works. As the rock in these storage areas has not been.placed. to any specification then the rocks are unstable and likely to move and trap construction workers or the public climbing on them. Where there is the possibility of temporary instability, this should be designed out of the construction process.

ORDNANCE

A significant amount of unexploded ordnance lies in the waters surrounding the UK as well as other parts of the world. The ordnance has either been deliberately dumped in specific locations, when sea dumping was a recognised and accepted means of disposal, or as a direct result of military/ wartime activity. Another potential source of explosives arises from possible charges which have been left unexploded in rock from quarrying or underwater rock removal sites.

The end result is that unexploded ordnance presents a risk to certain maritime activities. In addition to unexploded ordnance, there is a vast amount of scrap metal and non-hazardous waste that has also been deposited in the water. This can be misidentified as ordnance with the resultant delays to projects. Ordnance is designed to function in a pre-determined way. However, much of the ordnance to be found subsea has been dropped in an armed condition or has been fired and failed to function. In many cases the ordnance has deteriorated and safety features can no longer perform the function for which they were designed.

There is, therefore, a risk that unexploded ordnance can be caused to explode by direct impact or shock. Typically dredging, pile driving and other maritime intrusive engineering may provide sufficient energy to cause an item to function. Where practicable/feasible a threat assessment and underwater survey should be carried out to ascertain the presence or otherwise of unexploded ordnance in an area. If a proactive approach is not possible, reactive measures need to be explored. Some companies can provide Explosive Ordnance Safety and Awareness Briefings to all personnel working on a site as part of the overall site induction training and health and safety seminars. They can also provide services to deal with or remove ordnance which may reduce downtime on site. The assessment process is an imperfect science and clients find it difficult to find a balance between undertaking additional studies and managing the residual risk.

SUBSEA SERVICES

The existence of subsea pipelines and cables, especially electricity cables, oil and gas pipelines can pose hazards for coastal and maritime engineering, especially as it is more difficult to undertake intrusive and positive identification of the pipeline position below the seabed than it is for land-based operations. Marine geophysical survey techniques can usually be used to locate such services. Sonar or multi-beam bathymetric surveys can map pipelines lying on the seabed. A magnetometer survey can be utilised to determine the position of buried steel pipelines. If a depth of burial is required, or if the pipeline is constructed of non ferrous material then a seismic technique using a pinger or boomer sound source will be required. Detecting cables can be more difficult due to their smaller size.

A surface laid cable should be detectable with a high resolution sonar survey. Buried cables may be located with a magnetometer if the cable carries electrical current or has steel armouring. Otherwise specialised cable tracking techniques would need to be employed. For further information, it is recommended that the user contact a specialist marine survey company to advise on the best option to use.

TIMING OF THE WORKS

Coastal engineering tender documents issued by clients may have a contract duration specified, but rarely is an exact commencement date for the works given. The typical requirement for three months validity for tenders reflects a corresponding uncertainty in the commencement date. As a result, the weather statistics during the contract may reflect a more stormy or less stormy period than anticipated (especially if the contract is only to be of six to nine months duration). Uncertainty about the commencement date may be controllable by the client. Often lack of confidence in the availability and timing of funding is cited as a reason for not fixing the commencement date. Sometimes commencement dates are expressed in the.as soon as possible. category. Valid reasons for this may relate to the urgency of the project. However.urgency. can arise for a number of reasons. For example, this may arise due to: the need to make use of particular plant as soon as available; risk of failure of existing works; risk of loss of funds for the project (*e.g.* due to the approach of the end of the financial year for public sector clients) and political urgency. Sometimes the contract is timed to run through the winter period.

While there may be valid reasons for this, such as environmental concerns, concerns about damaging trade or tourism over the summer period, or more ready availability of plant and labour, fixing the timing in the winter period automatically increases the hazards to construction. However, there may be marginal benefits in carrying out the work during the winter period in terms of hazards to the public since there may be fewer people around the site in this period. If the contract timing can be made as flexible as possible then the contractor can maximise plant efficiently and avoid undesirable weather seasons.

5

Operational Plant Engineering Issues

Plant used in coastal construction works fall into three categories:

- Land-based plant, including earth-moving plant
- Marine plant
- Other specialist plant.

The various types of plant and their limitations are discussed below. It is important to understand that much of this equipment is not specifically designed to operate under the conditions it is likely to meet in a coastal/ maritime project.

This.misuse. means that the safety margins built into the plant by the manufacturers, applicable to the normal field of operation, will to some degree be eroded. Some equipment has been specially modified to help it meet these exceptional conditions, but often still remains at greater risk than in its original design environment.

LAND-BASED PLANT

Land-based plant is used extensively for construction of coastal works, both in transporting materials to and around construction sites, and in placing materials into their final position. The relatively large tidal range around the UK coast makes this possible for beach-based schemes, where work can be carried out using land-based plant in the intertidal area. Land-based plant can also be used for construction where the original levels are below low water, by making use of end-tipping of material to secure a working platform or by operating the plant from other temporary platforms or floating equipment (barges, pontoons, etc.).

Most conventional land-based contractors equipment can be employed on the shore-based aspects of coastal and maritime works. Normal safety risk assessment applies to the use of this equipment, but some items of plant and their use require specific site-based consideration. An example of this is the use of dump trucks on beaches. This particular equipment is susceptible to variable ground conditions and can easily overturn if a patch of soft sand is hit.

EARTH-MOVING PLANT FOR ROCK MATERIALS, BEACH NOURISHMENT AND FLOOD EMBANKMENTS

Relevant plant for rock materials and beach nourishment includes most types of earth-moving equipment. The more popular types are those which can operate readily on a beach, or near or on a breakwater. Hydraulic excavators are generally tracked rather than wheeled and may be fitted with a hydraulic bucket, Bofors grab or orange-peel grab. For moving smaller rock around on site, articulated steel-bodied dump trucks are popular, being able to accommodate rapidly changing ground levels. However, rigid dump trucks are also used, particularly for large pieces of rock armour. Hydraulic excavators fitted with grabs often load such material. Conventional earth-moving plant are used in the construction of embankments, and are similar to those described above but with the addition of bulldozers and rollers. There may also be the need to use craneage for long-reach activities, such as the installation of sheet-pile cut off walls, or draglines for trimming slopes.

PLANT FOR HARD DEFENCES

As well as earth-moving equipment, the construction of hard defences requires additional heavy plant in the form of cranes, which are required to lift or suspend objects such as:

- piles and piling hammers or vibrators for steel sheet piling
- precast concrete elements such as precast concrete beams, steps and armour units. (Lighter elements can be lifted by hydraulic machinery as long as the required reach is not too large)
- formwork for in situ concrete during fixing in place
- concrete skips (where not pumping in situ concrete).

PLANT FOR CONSTRUCTION OF MARINE-PILED JETTIES AND DOLPHINS

Installation of the tubular steel piles for these structures typically uses heavy piling hammers located either on A-frames on large floating pontoons, or suspended from ringer cranes mounted on jack-up barges. Construction often requires the use of some form of temporary steel.leader. framework to hold the pile in position until it has been driven sufficiently far into the bed to achieve an acceptable degree of rigidity. Driving such piles into the seabed often requires pre-drilling and the large diameter drills required also need craneage to move them around.

Marine construction work often makes extensive use of precast concrete pilecaps, beams and slab elements in order to reduce the cost and safety issues associated with over-water formwork. However, some in situ concrete production is inevitable, unless the deck or the deck joints are fabricated out of steelwork and rely on welded or bolted connections. Commonly, steel deck frames are restricted to small areas, such as in mooring or berthing dolphins. Larger areas are normally formed with concrete and have some kind of in situ concrete topping. The in situ concrete can be provided from a floating batching plant. If mixed on land, the wet concrete will be shipped out in skips or rotating mixer drums and placed in its final position using marine cranes or land-based cranes located on already complete/ partially complete parts of the structure. Pumping of concrete may also be feasible, but only over relatively short distances.

MARINE PLANT

Like land-based plant, marine plant is also used for both transport and placing. They are subject to access limitations dictated by the water depth available for navigation, a depth which is constrained by both variations in water level and by wave action. International legislation covers the safety of marine plant with the International Maritime Organisation (IMO) providing the overarching governance. A number of the problems that arise with marine plant originate from the fact that, while each captain is responsible for the safety of their own vessel as a first priority, they may be inexperienced in managing it in the inshore environment. The particular wave, current and water depth phenomena of that environment have an ability to punish even small errors both quickly and severely. However, inexperience of this environment can be mitigated by employing a captain or beachmaster with the necessary experience of operating in these conditions. This beach master is charged with the responsibility of ensuring that operations are carried out in a safe and workmanlike manner, and that no skipper inadvertently puts either their own or another vessel at jeopardy. Accident rates both in terms of death and injury, as well as damage to equipment, are significantly lower on contracts where this strategy has been employed. Marine plant can be divided into six main categories:

- work boats
- ships (including any self-propelled module)
- non self-propelled barges and pontoons (*i.e.* anything which requires tug assistance to manoeuvre)
- marine craneage
- jack-up platforms
- safety/rescue boats.

WORK BOATS

A work boat is a general term which applies to boats used to undertake a wide range of functions including the support of marine construction work. Many are standard in design but there are many variations. Tugs are specialised push/pull vessels, ranging from small vessels designed for use in harbours and inland waterways to much larger ocean-going vessels. A common factor to nearly all tugs is that they are deep hulled for their size, and often require more draught than the equipment they are moving. Multicats, powerful motorised barge/pontoons complete with a hold and engine room, bridge, clear working deck and some form of mechanical lifting equipment, are frequently employed in most marine/harbour works. Work boats can be used as a shuttle vessel, tug, ferry or floating working platform. They tend to be extremely flexible and can carry out work which would otherwise tie up major items of marine plant.

A code of practice has been developed by a number of organisations to assess and set standards of safety and protection for all on board. Moth provides an easy to use guide to this code of practice for vessels under 24 m in length. The.Code of Practice for the Safety of Small Work Boats and Pilot Boats. does not apply to vessels operating on inland waters where such waters form part of a river or are completely self-contained such as a lake or reservoir. In such circumstances, further advice should be sought from the river and/or water authority or the local authority regarding any local requirements that might apply. A new.Code of Practice for Small Commercial Craft. is currently under preparation by the MCA. This new code has been developed for application to UK vessels of up to 24 metres Load Line length that are engaged at sea in activities on a commercial basis, that carry cargo and/or not more than 12 passengers, or provide a service in which neither cargo nor passengers are carried, or are UK pilot boats. This new code of practice will supersede the following four Codes of Practice:

- The Safety of Small Commercial Motor Vessels (Yellow Code).
- The Safety of Small Commercial Sailing Vessels (Blue Code).
- The Safety of Small Work Boats and Pilot Boats (Brown Code).
- The Safety of Small Vessels in Commercial Use for Sport or Pleasure Operating from a Nominated Departure Point (NDP) (Red Code).

Publication of this new code was scheduled for the autumn of 2004. Moth provides guidance based on the legislation on the protection of personnel on work boats with respect to the following areas:

- Deckhouses
- Bulwarks, guard rails and handrails
- Safety harnesses
- Life-saving apparatus
- Surface of working decks
- Retrieval of persons from the water
- Personal clothing
- Noise.

DREDGERS

Dredgers are used for excavating the seabed and either dumping the material at sea or pumping it ashore, for land reclamation or beach nourishment. Their hulls are derived from ocean-going ship design, which are meant to operate in harbours or the open sea. They are built on the assumption that they will operate with sufficient water under their keels. Side stone dumpers are specialist vessels designed to place smaller armour stone very accurately underwater. To this end they are equipped with very sophisticated navigation and manoeuvring equipment. This ability is largely wasted if they are used simply to dump large rocks on a beach for re-handling. Apart from a few small vessels carrying under 1000 t at a time their draught requirements are from 3.5 m upwards.

NON SELF-PROPELLED BARGES AND PONTOONS

For shallow water work, dredging can also be carried out with a spud-leg motorised barge equipped with a slew ring mounted backacter. This equipment, used with a split hopper barge, is well suited to shallow or confined waters to which a conventional dredger cannot get. The spud legs make them particularly stable during operations. Barges are broadly boat-shaped, but often need a tug to move them around although some are self-propelled. Many are adapted for dumping materials at sea in conjunction with static dredgers, and are also commonly used to deliver beach nourishment or for rock dumping. Other uses of barges include acting as platforms for piling frames and light equipment, floating stockpiles, and materials delivery.

Towed equipment needs more sea room than self-propelled vessels, particularly under windy conditions, and may be draught limited by their tug. Simple pontoons are also extensively used on marine construction sites as transport for temporary and permanent materials and as working platforms for lifting and/or pile driving. Uniflotes or equivalent can also be joined together to make platforms of any size and shape to fit around or between construction works.

Pontoons are basically floating steel boxes used as floating platforms for a wide variety of uses. Some are fitted with.spuds. to hold them in position without the need for anchors (spuds should be distinguished from jack-up legs. spuds are a positioning tool and cannot be used to lift the equipment above the water). Sizes of pontoons vary enormously from Uniflotes, working either singly or in assemblies, to a 20 000 t rock barge. Some are dedicated to permanent equipment such as large cranes; most are modified for some degree of dedication even if only on a job-by-job basis. Some pontoons are designed to sit on the seabed or beach. However, care should be taken as a small error in procedure can lead to very expensive repairs or even a write-off..Beaching. has therefore become unpopular with insurers and as a result is now much less popular. The only pontoon specifically built for carrying rock operating in European waters is a 22 000 t unit; all others used are modified from other purposes. Some, mostly continental, units are designed for inshore work, but not for the severe conditions so often found in UK waters. The majority of serious accidents to non self-propelled equipment occur when they are trying to ride out bad weather close to a lee shore and they suffer some form of failure of their anchoring system. It is good practice therefore to have sufficient towing units on station, which can either take them to shelter, or give them extra sea room in the event of an on-shore storm. Provisions for edge barriers preventing plant from going over the edge when operating on the barge need to be considered and managed.

JACK-UP PLANT

Jack-up platforms are pontoons fitted with legs and lifting gear which bear on the bottom and allow them to be lifted above the water to work. In other words they manoeuvre as marine plant and work using the.land. to support them. In general they can carry a much higher deck loading when jacked up than when afloat. They are at their most vulnerable when manoeuvring and jacking up as they are normally reliant on tugs or anchors to move, and it is rarely possible to inspect the seabed in sufficient detail to be certain that it is capable of taking the load of the legs. There have been a number of accidents caused by failure as a leg is being loaded up leading to tilting, leg breakage and consequent overturning: more rarely, this can happen in service when the bed becomes overstressed with time. Another known form of failure arises from the legs not being long enough to keep the pontoon above storm waves, and it is quite literally lifted off them. However, in general, they have a good safety record, and can be confidently used in situations where other types of pontoon would be a severe hazard.

SAFETY BOAT/RESCUE BOAT

Safety boats can include:

- guard vessels in the vicinity of the works
- standby boats in the vicinity of the works.

A *rescue* boat is different and is a boat only put into action when an incident has occurred (*e.g.* RNLI vessels are rescue boats). Many contractors considered that safety boats often provided a false sense of security. Their benefit in calm sea conditions was considered marginal while they often could not be launched or operate safely in severe swell conditions when they would provide the most benefit. Their very existence may put the safety boat operatives in an unsafe environment. Those consulted considered that a risk assessment should be undertaken and, where possible, mitigation measures should be put in to avoid operatives falling in the water in the first place rather than providing a safety boat. Selecting the most appropriate boat is very important. The RNLI may be able to provide guidance on selection. Issues to consider are given below.

- In open waters the choice is open. Often a fast pilot cutter is the most viable option as it has a low freeboard and offers some protection from the elements to the crew if they are on standby.
- In the surf zone a RHIB is the only viable alternative. However, it is not usually viable to have a RHIB in the water on standby as there may be more of a risk of losing crew than a man overboard from the works. Instead it may be more appropriate to have a RHIB on a very quick launching davit if working from a jack-up or land structure (jetty).

It may be possible to provide a mooring if working from a floating vessel. The difference is that the tide makes mooring against a land structure more difficult and can result in damage to the boat. Often an experienced crew can be recruited from local fishermen. Their local knowledge of land and sea can be beneficial and initial training can be minimal. If possible, the contractor should choose those who are serving with the local lifeboat as their training will be beneficial to the project.

OTHER PLANT

Landing craft exist world-wide as a variety of forms of vessel usually designed to meet military needs in the first place. which have a capacity to beach themselves and offload via some sort of ramp. In UK waters these range from small units, meant to carry personnel, to ocean-going ships with bow doors and ramps for offloading heavy plant. Their common attributes are very shallow draught (2 m or less maximum) and a ramp or ramps. They tend to be much more seaworthy than they feel, and the biggest danger to them is underwater obstructions in the beaching area. For this reason, it is good practice to establish a permanent landing area for them which is known to be free of obstructions at any state of tide, and which offers a good flat surface on which they may rest while discharging without risk of.breaking their backs.. (Such landing area criteria apply equally to situations where any vessel or pontoon may be beached.)

The majority of land-based equipment with a marine capacity started life as a military vehicle. As nearly every government has different ideas as to their own military requirements, there are some very odd vehicles about. They may be divided into two types, wheeled and tracked, after which the possibilities for variation are endless. Intending users should ensure that they fully understand the exact capabilities and limitations of any vehicle, as similarities with other vehicles of which they may have previous experience may be deceptive.

6

Code of Practice for Marine

SAFETY PLANNING

SAFETY PLANNING REQUIREMENTS

This Code of Practice provides guidelines on safety planning for site operations. Most of this information can also be found in the Global Marine Systems Project Safety Plan, but this CoP is angled more to the site personnel's point of view and it covers new build projects, large service projects, crane projects and service trips.

SCOPE

These requirements apply to all Marine and Cranes employees and contractors when working on site. If national legislation or customer requirements impose a higher standard, then the higher standard shall be followed. In such cases this shall be documented.

HAZARDS AND RISKS

Environment, Health and Safety (EHS) aspects for site works shall be identified, evaluated and correct controls shall be put into place in accordance with the hierarchy of controls. This has to be done by all MARINE personnel and contractors. In order to be able to do this properly, persons in different roles need suitable EHS competence. The following personnel are especially important while securing proper EHS for site operations: Project Manager (PM):

- The Project Safety Plan is established and implemented during the project start-up and is updated as relevant. is overall responsible for EHS in the project and shall ensure that:
- EHS risks are identified, evaluated and actions taken as necessary. PM shall seek customer clarification in this matter, and also about the customer's safety rules and regulations.
- EHS planning, scheduling and documentation of activities to eliminate or minimize significant EHS risks.

- Establish Project Emergency Contact Plan.
- Give relevant ESE documents to site personnel, and ask for relevant EHS document from site personnel for monitoring of EHS performance.

Site Manager (SM):

- Site safety risks are assessed as required and actions taken as necessary. (when applicable) is responsible for EHS on site related to MARINE personnel and contractors and is accountable to PM. SM shall ensure that:
- Unsafe activities are immediately stopped and safe operations restored.
- Safety Data Sheet (SDS) are available for all MARINE supplied chemicals.
- Coordinates/communicates tasks with yard and contractors, *e.g.* scheduling tasks in safe order.
- MARINE personnel and contractors get EHS briefing on site.
- All MARINE personnel and contractors on site follow the Project Safety Plan.
- EHS incidents and accidents are reported.
- Safety Tours at site are performed.

Manager selecting persons to site:

- Site personnel (employees and contractors) are competent, trained and formally authorized to work safely at site. (*e.g.* Commissioning Manager) shall ensure that:
  - Personal Protective Equipment (PPE), tools and other equipment are available and functional/ calibrated. Sales Manager
  - MARINE safety requirements are communicated to the customer. shall ensure that:
  - If project is executed at new yard, safety advisor is consulted regarding risks prior tendering.
  - Safety costs are taken into consideration in the sales proposal.

Site Personnel:

- Instructions are followed and selected controls implemented in general shall ensure that:
- MARINE Stop Take Five philosophy is followed.

OPERATIONAL CONTROLS

Safety be shall part of preparation and execution of all site works. It shall be in the agenda of project meetings as well as start-up meetings for the whole project (internal and with customer), for commissioning meetings (internal and with customer) and it shall be part of execution. Suitable level corrective action shall be taken on reported near misses, unsafe conditions and behaviors based on risk assessment. All site personnel are responsible to:

- Familiarize themselves with MARINE Project Safety Plan, including participation in customer safety orientation (if applicable).
- Follow the prevailing instructions, safety rules and regulations, including the BU Marine and Cranes Codes of Practices.
- Check regularly that PPE and tools are functional and in good order. Report malfunctioning PPE and tools to supervisor.
- Report near miss incidents and accidents, unsafe conditions and unsafe behaviors to site/project manager and line manager.
- Perform JOR, SJA/JSA/JRA, SOT and tool box meeting when required.
- Follow MARINE Stop Take Five philosophy.

Safety Observation Tours (SOT)

Safety Observation Tours shall be performed periodically by Project manager/Site Manager. SOT is required for all projects with an MARINE Site Manager. For projects without MARINE Site Manager PM shall perform SOT during site visits.

Job Observation Report (Applicable for NO)

Before the site work can start, the Commissioning/ Service Engineer has to make sure that the work site is ready. Job Observation Report (JOR) has to be filled out by the Commissioning/ Service Engineer before work is started, and again if he/she has been away from the site for more than one week. This is to ensure that required level of EHS is in place before the work takes place. Completed JOR is to be sent to MARINE Project Manager/ Site Manager for archive in project file. If there are any deviations or actions required, MARINE Project Manager/ Site Manager (or Client site responsible) is to be notified immediately. If any hazardous conditions are revealed, the work shall not be started before SJA is performed and/ or necessary actions carried out.

Safe Job Analysis/ Job Safe Analysis/ Job Risk Analysis (abbreviated here SJA)

A "Safe Job analysis" (SJA) shall be performed, if an unfamiliar task is identified *i.e.* it is not covered by existing controls (*e.g.* procedures or instructions) or existing controls are not considered to be adequate (Residual risk in Risk Assessment remains too high). In the SJA, all conditions and activities in the specific working area shall be considered. The procedure describing in detail securing of safety for personnel, material and environment, are to be worked out together with involved personnel. Completed SJA are to be signed by the persons that have worked out the document, sent to

MARINE Project Manager/ Site Manager, Client site responsible (if applicable) and stored in project files by MARINE Project Manager. All relevant personnel shall be informed.

Tool box meeting at Site incl. JESA (Applicable for CN)

The purpose of the Toolbox meeting is to achieve an exchange of information and ideas on health, safety and environmental matters. The Project Site Manager/ Site leader shall be responsible for conducting Toolbox meetings for all employees in their respective work areas including contractors. The Project Site Manager or nominee shall ensure a record of Toolbox Meetings is completed for each safety talk given, including topics and names of employees attending. A copy of the Toolbox Talk and relevant open actions will be forwarded to the Project Manager on completion for review and follow up.

TRAINING & COMPETENCE

The Health and Safety (H&S) of MARINE personnel when working on site will be determined to a large extent on their competence to undertake the work. This code of practice describes the minimum training and competence requirements for Marine and Cranes personnel working in a site environment.

SCOPE

This COP applies to all Marine and Cranes employees and contractors working on customer sites. If national legislation or customer requirements impose a higher standard, then the higher standard shall be followed. In such cases this shall be documented.

HAZARDS AND RISKS

Without suitable health and safety competence persons working on site will not be able to identify the hazards correctly, and therefore will not be capable of evaluating the risk and putting into place the correct controls in accordance with the hierarchy of controls. Risk is the product of hazard severity and the probability of the occurrence of harm. The latter will be determined by a number of factors and a fundamental factor that will have a significant effect on the risk of injury from a particular task or activity will be the degree of training and competence. If MARINE personnel are not trained or competent then the risk of injury will increase

OPERATIONAL CONTROLS

This code of practice defines the minimum occupational health and safety training/competence requirements for all site personnel. Additional training/ competences can be/are required in some specific work tasks. All personnel working in a customer site environment shall be skilled for the work in question and instructed in the health and safety contents and requirements set out in these Marine and Cranes codes of practice. All persons shall also receive suitable training and/or instruction on the personal protective equipment (PPE):

- Correct use of the PPE that they have been issued with and what not to do;
- How to fit RPE (Respiratory Protective Equipment) correctly;
- How to obtain replacements.
- Emergency arrangements where PPE may be needed.

All persons who attend site shall receive a suitable site induction briefing to ensure that they are fully familiar with (accountable: Site Manager. If not in place the Project Manager or Customer):

- General site layout
- Access points to the site;
- Safe walking routes;
- Location of welfare facilities including toilets, washing, canteen, drinking water facilities;
- Location/arrangements regarding medical help;
- Site first aid arrangements;
- Emergency evacuation arrangements and assembly points;
- Site OHS rules including PPE requirements;
- Emergency contact phone numbers (including customers' representative on site).

The following valid safety trainings are required for persons involved in mechanical work

- First aid (preferred for all, in minimum as defined in CoP 8 "first aid and emergency medical")
- Basics fire training (*e.g.* use of fire extinguisher)
- Basics on Electrical safety including Seven Steps approach and safety distances
- Hot work license (preferred, especially in ship building environment)
- Azipod Confined Space (if working inside Azipod unit)
- Scaffolding inspection (min. one person at site, if work involves working on scaffoldings)

The following valid safety training courses are required for electrical work

- The above-mentioned courses
- Electrical Safety Training (electricians level, this replaces Basic training *i.e.* point c) above)
- Live electrical work safety training

Project and site managers must have role specific valid health and safety training.

RAINING AND COMPETENCE

Every project team member must be familiar with these requirements prior starting any work on site.

MONITORING AND CHECKING

The LBU manager is responsible to secure that managers selecting persons to sites are following these requirements. Managers selecting person for site works shall be responsible for ensuring that all persons who are required to carry out work on site have the required trainings and competence.

DOCUMENTATION & RECORDS

Managers selecting persons to sites must be able to show records proving that the minimum occupational health and safety requirements for the selected person(s) are met. Site managers must be able to show records that proper site briefing has taken place.

TRAVELLING REQUIREMENTS

The requirement for international travel is an integral part of the Marine and Cranes business requirements and travel requirements need to be planned well in advance. This will save time and reduce any personal stress that might arise as a result of having to organize things in a hurry. This Code of Practice (COP) has been prepared to summarize the key requirements in respect of travelling to site.

SCOPE

These requirements apply to all Marine and Cranes employees and contractors. If the local regulatory or customer requirements impose a higher standard, then the higher standard shall be followed. In such cases this shall be documented.

HAZARDS AND RISKS

Hazards and risks will vary and will depend on travel arrangements and the destination country. The following need to be considered and evaluated:

- Travel and medical insurance,
- visa requirements,
- declaration of work to be performed
- transportation and tickets,
- immigration and customs documentation,
- accommodation,
- local contacts including MARINE host country organization,
- diseases – vaccination,

- medication and medical treatment,
- unusual blood type, transportation in country of destination,
- invitation letter by customer,
- working conditions including working hours,
- required personal protective equipment(PPE).

For high risk countries identified in the MARINE Security Threat Map the country specific hazards and risks are described within the Travel Code of Conduct for the country and authorization for travel may be required from the Host Country Manager. For the identified Water Threat Level areas the specific hazards and risks are documented in the Ship Security Plan.

OPERATIONAL CONTROLS

The following shall be covered as a minimum:

- Approval is required in all cases by line manager prior to travel.
- Travel to be arranged through an MARINE approved travel agency and details will be entered into the travel tracker system automatically. This applies also when travelling to countries with threat level 3.
- In those cases where travel is arranged by a non MARINE travel agency then the details shall be entered manually.
- Hotel booking arranged through Hotelzon or an approved travel agency
- Travel checklist filled in, if applicable

TRAINING AND COMPETENCE

MARINE Travel policy, MARINE general advice for travelling and working internationally, and Emergency procedures shall have been communicated to Marine and Cranes employees and/ or contractors prior to travelling.

MONITORING AND CHECKING

Communication arrangements and travel plan shall be available to the LBU responsible manager/ unit. All incidents, unsafe conditions, unsafe behaviors and near misses shall be reported to the applicable database.

DOCUMENTATION AND RECORDS

Travel documents and plan, management approval, hotel booking and contact persons shall be made available by the LBU responsible unit, e g in the travel database.

HIGH RISK COUNTRIES

As a global company, MARINE works in almost a 100 countries. The political situation in some countries and in certain international waters, where

MARINE might operate could represent a significant risk to employees working there. MARINE has well established travel security procedures to identify those countries where threats are higher and also to assist in monitoring the status of travelers and to receive updated advice. This Code of Practice has been developed to assist Marine and Cranes personnel in complying with these requirements.

SCOPE

This COP applies to all Marine and Cranes employees and contractors. If the local regulatory or customer requirements impose a higher standard, then the higher standard shall be followed. In such cases this shall be documented.

HAZARDS AND RISKS

Hazards and risks will very much depend on the required travel arrangements and the country destination. Typical risks that could be encountered include:

- social unrest;
- kidnapping & extortion;
- theft;
- armed robbery;
- piracy;
- terrorism;
- transport and traffic situation;
- natural disaster.

Countries that are identified in the MARINE Security Threat Map will have the country specific hazards and risks described in the Travel Code of Conduct for the country. For the identified Water Threat Level areas the specific hazards and risks are documented in the Ship Security Plan. These procedures are regularly updated and general advice is also provided on international travel and global risks.

OPERATIONAL CONTROLS

The following control measures shall be applied as a minimum:

- The employee who will be traveling shall consult the MARINE Security Threat Map for Threat Level on land and in waters and check on current requirements;
- Ensure that prior to departure that the traveler is suitable to travel to high risk rated areas, and also willing and ready to go;
- The person who will be traveling shall undergo medical screening to ensure that he is in good physical fitness;
- For travel to countries threat level 3 and above, approval is required

by LBU Manager and visit should be registered with the host country by ensuring that the ticket is purchased through an MARINE travel agency and details put on the travel tracker system. This is done automatically but for those arrangements which are not made through an MARINE approved travel agency then the details shall be entered manually.

- For travel to countries where the threat level is 2 then approval is required by LBU Manager and details to be entered into the TIS system. Prior approval is required from Local Division Manager with copies to host Country Manager and MARINE Group Travel.
- For travel in high risk countries (threat level 1) additional authorization by the local Division Manager (LDM), Country Manager and MARINE Group Travel is required prior to travel;
- Travel Information Sheet (TIS) shall be completed and submitted five days prior to travel
- Ship Security Plan (SSP) completed and submitted prior to travel if in Threat Level waters
- Automated Travel Advisory (ATA) will be sent to the traveler upon booking a trip and will contain advice and recommendations regarding medical and security advice for the selected country.
- Traveler shall normally be briefed about the security situation and relevant instructions, usually by the Security Manager International Operations, using the general security brief and selected country security brief

TRAINING AND COMPETENCE

MARINE Travel policy, MARINE general advice for travelling and working internationally, Travel code of conduct for the Treat Level area and Emergency procedures shall be communicated to Marine and Cranes employees and/ or contractors prior to travelling.

MONITORING AND CHECKING

- The LBU responsible manager/ unit shall ensure that the communication arrangements and travel plan is available in the Travel Information Sheet database.
- It is important that the traveler notifies the host country organization of his presence in the country. All incidents, unsafe conditions, unsafe behaviors and near misses shall be reported to the applicable database.

DOCUMENTATION AND RECORDS

- MARINE uses a travel tracking system that enables quick access in order to identify who is in a particular affected area in case of a crisis or incident. It is therefore a record.

- For high risk countries (threat level 1 and 2) the TIS system is used in addition.

DRIVING & TRANSPORT SAFETY

Road travel is one of the major areas for accidents and incidents. This code of practice (COP) has been prepared to provide practical guidance on best practice to be followed in respect of driving and road safety. The objective is to promote travel safety and security, ensure proper planning, efficiency and reasonable costs as well as harmonize travel-related practices.

HAZARDS AND RISKS

Road incidents often cause more deaths and serious injuries to employees and contractors than incidents on site. Unsafe vehicles, violations against regulations and speed limits, fatigue, etc are examples of hazards and risks. Travel is an important part of work undertaken by BU Marine and Cranes, and within MARINE Group driving is identified as one of the high risk tasks for our personnel.

OPERATIONAL CONTROLS

Risk Avoidance & risk reduction

- Use videoconferencing, same time meetings and teleconferences instead of traveling whenever practical;
- Plan early, if possible. Advance planning reduces risks and makes it possible also to find the lowest prices;
- Preferably pre-book transportation prior to travel via our travel agency partners, or local MARINE colleagues;
- If booking a vehicle at the location, ask assistance from local MARINE colleagues. Consider language barrier as it is common that there are often only very few English speaking persons in the leasing companies and they can usually speak only some simple words;
- Use MARINE preferred rental companies (agreement between MARINE Group and the rental company) in case vehicles are leased, rented or hired. This should ensure that vehicle meets safety standards;
- Do not drive vehicles rented by colleagues, unless you are registered as a co-driver on rental papers;
- Inspect the vehicle prior acceptance especially the existence of seat belts, condition of tires (wearing surface, air pressure, in snow conditions winter tires), at least two airbags "driver and passenger",

head rests at seats in the front and at back, ABS and power steering and first aid kit, air-conditioning in hot environment. The vehicle should not be older than 3 years;

- When booking vehicle with a driver, clearly emphasize requirement of seat belts in the back seat and need for safe driving practices each time to raise the attention! Whenever possible, travel in the back seat. Avoid 'spot' or hire direct taxis in high road risk countries.

Risk control

- Use seatbelt on each occasion and travel only in vehicles specifically designed for passenger use;
- If the vehicle is also transporting goods, ensure that the load is secured properly;
- Comply with all national road laws, and site speed limits;
- Do not use hand-held cell (mobile) phones, or radios when driving;
- Ensure the manufacturers limit for passengers, and load to be carried are not exceeded and follow procedures for safe operation and maintenance of the vehicle;
- Report all motor vehicle incidents and near misses. Take photos if possible and include details of those involved;
- Do not use motorcycles for business travel;
- Do not under any circumstances drive under the influence of alcohol or drugs, or when suffering from fatigue;
- Always assess your capability for driving, especially after completing long haul flights etc. If any doubts, use alternatives *e.g.* host pickup, appropriate taxi or public transportation;
- One common root cause of road incidents is fatigue. It is recommended that adequate rest is taken on long journeys and that a short break, say of at least 15 minutes, is taken after about 2 hours driving with a longer break, of about 60 minutes, after 4 hours;
- Since so many factors affect fatigue, no firm limits in terms of mileage or time are possible to set, except perhaps for very specific situations. Examples of recommended maximum distances set for some specific situations are 400 km (250 miles) for driving on generally good roads and 600 km (375 miles) for mainly motorway driving. However, any such recommendations should be regarded as exceptional journeys, not the norm. Mileage limits should be regarded as indicative and not rigid otherwise unreasonable situations could arise, such as an overnight stop a short distance from the destination or the use of cross country or town centre routes instead of an easier but longer motorway journey.

TRAINING AND COMPETENCE

- Ensure that you are suitably trained and licensed to operate the class of vehicle to be used;
- Employees who drive more than 30,000 km for business per year, or who are required to work in an area with a High Risk Level status shall attend a suitable advanced driving course.

MONITORING AND CHECKING

- The supervisor is responsible for ensuring that persons who are required to travel are aware of this Code of Practice;
- Individuals who are traveling are responsible for following the requirements of this Code of Practice;
- Driving and road safety practices shall be part of OHS audits and reviews conducted by the LBU.
- All incidents, unsafe conditions, unsafe behaviors and near misses shall be reported to the applicable database.

SITE FAMILIARIZATION

A lack of hazard awareness, or working in environments which are unfamiliar can increase the risk of injury purely because persons working on the site do not have the site specific knowledge about the hazards and risks on site, as well as the control measures and procedures in place to deal with those risks.

This code of practice (COP) therefore sets out some basic requirements to be followed when planning any visit to a shipyard or port to ensure that all MARINE staff and any contractors are aware of the location, the site specific hazards and the general arrangements on site for controlling those risks.

SCOPE

This COP applies to all Marine and Cranes employees and contractors, or work sites which have a nominated MARINE Site manager or responsible person (and partially to personnel working for small and/or short term site works, such as service visits). If national legislation or customer requirements impose a higher standard, then the higher standard shall be followed. In such cases this shall be documented.

HAZARDS AND RISKS

Working within any unfamiliar environment will increase health and safety risks owing to the lack of knowledge about any site specific hazards and any site specific rules, practices or procedures, or any prohibitions that are applicable.

OPERATIONAL CONTROLS

Planning

- Much can be done before arriving at site by ensuring that prior to any contract being agreed suitable enquiries are made with the customer to ensure that details of the site, including its layout, details of site specific hazards and any site safety rules that apply are obtained.
- The information obtained shall be incorporated into the H&S plan for the project/contract which shall be prepared during the tender/ contract phase.

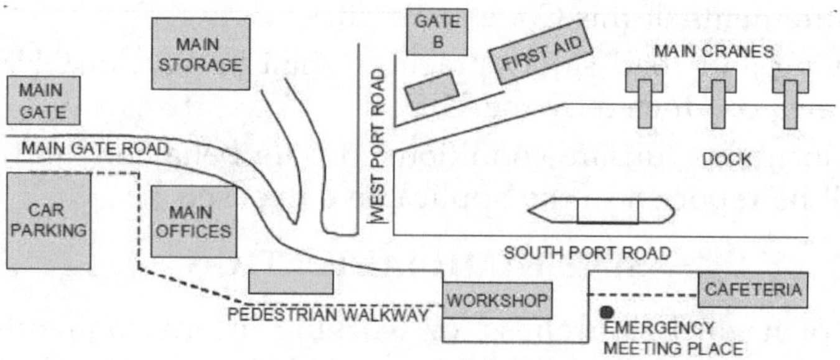

Fig. Illustration of Site Layout

The information that shall be obtained shall include

- a site map and details of any safe walking routes, prohibited areas and location of the facilities including welfare, first aid etc;
- Details of the customers' on site organization and contact person or a contact for the yard or port;
- Details of the site EHS rules which should include information in respect of any or all of the following:
  - Prohibitions; *e.g.* alcohol and drugs
  - Requirements for personal protective equipment on site
  - Radio practices, *e.g.* frequencies
  - Announcement practices
  - Lock out/ tag out – practices
  - Requirements for work permits *e.g.* hot work, electrical isolation etc
  - Fire and emergency procedures including emergency telephone numbers for fire, police, ambulance and emergency contact for the yard or part facility;
  - Waste treatment arrangements
  - Site security requirements
  - Vehicle parking requirements

- Storage and lay down areas
- Incident reporting requirements including near miss and hazard reporting requirements
- Disciplinary matters on site.

ON ARRIVAL AT SITE

- On arrival at site all MARINE employees and contractors shall report to the security gate and sign in and obtain their site ID if required;
- The MARINE supervisor or site manager shall ensure that MARINE employees and contractors are given a site briefing or induction regarding the site EHS rules and the emergency arrangements on site.
- Where the above is not provided then the MARINE supervisor or site manager shall contact the customers' representative on site in order to obtain this information;
- The site briefing or induction shall if possible cover the following topics:
  - Site layout specifics including safe walking routes
  - Important locations; *e.g.* parking, gates, lunchroom, emergency meeting places, offices, welfare facilities and first aid station;
  - Emergency contact information including telephone numbers for ambulance, police, fire service and also the port, yard or vessel contacts;
  - Vessel or Crane specifics, *e.g.* evacuation routes, fire extinguishers.
- For small and/or short term works, such as service visits where the works are <2-3 days, the MARINE employee executing the work shall request a site escort from the Customer who is familiar with the site who can brief the MARINE employee on the site and its various facilities and any EHS requirements. Internal MARINE EHS requirements, such as emergency protocols, PPE etc. are not handled in this COP.

PRIOR TO STARTING WORK

Prior to starting work the MARINE supervisor or site manager shall carry out a briefing on:

- The scope of work to be carried out on site;
- Principal hazards and risks on site including presence and location of asbestos, radiation sources and in particular any non destructive testing on board the vessel; storage and use of hazardous chemicals etc;
- EHS (Environmental, Health & Safety) site requirements;

- Name and phone number of EHS responsible person on site or vessel;
- MARINE responsible person for commissioning of MARINE unit/ service;
- Name of responsible person on board the vessel for connection or switching
- Details of any other contractors who may be working within the same working area as MARINE and any risks that may affect MARINE;
- Location of general welfare arrangements including
  - Supply of drinking water
  - Lunchroom and messing facilities
  - Toilets and washing facilities
  - Changing facilities and storage of clothing
  - Storage of personal protective equipment
  - First aid and medical facilities.

TRAINING AND COMPETENCE

This COP shall be communicated to every project team member prior to starting any work on site.

MONITORING AND CHECKING

The MARINE supervisor or site manager is responsible for ensuring that all project team members have been familiarized with the site prior starting any work. This shall include contract personnel working on behalf of MARINE. Site familiarization shall be part of EHS audits and reviews conducted by the LBUs. All incidents, unsafe conditions, unsafe behaviors and near misses shall be reported to the applicable database.

DOCUMENTATION AND RECORDS

Site Familiarization shall be included as a part of the Project Safety Plan during project start-up phase and documented during the commissioning phase.

SAFETY SIGNS & SIGNALS

It is generally accepted that safety signs and signals are not always understood by employees and others. This code of practice has been prepared to provide general information to MARINE employees on signage that is used internationally to communicate such information clearly.

SCOPE

These requirements apply to all Marine and Cranes employees or contractors when working on site. If the local regulatory or customer requirements impose a higher standard, then the higher standard shall be followed. In such cases this shall be documented.

HAZARDS AND RISKS

It is an important safety measure that all MARINE personnel and MARINE's contractors on site are able to easily understand the safety signs. MARINE employees who do not understand the system of safety signs and signals so could result in them failing to recognize a known hazard or another danger such as a prohibited area or an action which in turn could lead to serious injuries or even death. This will also apply to contractors or in some cases the public *e.g.* visitors to site.

FIRST AID & EMERGENCY MEDICAL

It is essential that all Marine and Cranes personnel or contractors are aware how to act in cases where emergency medical treatment is required and/or first aid assistance is needed on site. This code of practice (COP) sets out basic requirements to be followed when working on customer sites.

SCOPE

This COP applies to all Marine and Cranes employees and contractors, or work sites which have a nominated MARINE Site manager/responsible person. If national legislation or customer requirements impose a higher standard, then the higher standard shall be followed. In such cases this shall be documented.

HAZARDS AND RISKS

Marine and Cranes work sites can often have poor accessibility for medical teams. This combined with imprecise location information of the injured person can increase the risks further. Failure to either provide for adequate first aid provision or emergency medical support is likely to increase the severity of outcome of any injury that may have been incurred.

OPERATIONAL CONTROLS

General

- All contracts or projects tendered for shall include suitable provision regarding the reasonable availability of emergency medical services. It is impossible to state what is reasonable, but at least the provisions must be such that a seriously injured person will be treated by a medical professional within one hour of any incident. This will normally be at hospital, or suitable trauma center. To achieve the response time for seriously injured persons at remote sites/locations

it may be necessary to make/secure additional medical provisions at site, such as
  - Ambulance or vehicle capable of taking full length stretcher
  - Availability of medical professional, and trauma facilities
  - Helicopter evacuation

The above must be established from the customer during the tender stage of the contract.

- As an absolute minimum requirement every work location shall have a first aid available within 5 minutes of incident.
- All projects and work areas shall have a first aid kit readily accessible so that the first aid can start within above mentioned 5 minutes of an incident.
  - For measure of best practice, and/or if it cannot be determined if adequate First Aid kit is readily available on site, MARINE shall provide their own First Aid kit to every project.
  - First aid kits shall be constructed of impervious material, dustproof and of sufficient size to store the required contents. They must be capable of being sealed and have a handle for emergency transport.
  - The exterior of the first aid kit must clearly identify it's purpose, for example - "First Aid".
  - Contents of the kits shall be suitable and sufficient for the site and shall be based on the hazards and risks involved *e.g.* if welding is carried out then treatment for burns must be considered.
  - Every first aid kit shall be kept fully stocked and maintained in a clean and hygienic condition.
- Emergency contacts information shall be posted visibly around site, and communicated to all workers. This shall include telephone/Radio contact name & numbers of
  - Medical professional
  - Ambulance
  - Hospital

Emergency medical care

Emergency medical care is the provision of skilled medical help at the scene of an accident, medical emergency, or during transport to hospital. It consists of recognition, resuscitation and stabilization of the seriously injured and it extends beyond the preservation of life to the prevention of complications and the relief of suffering.

EMERGENCIES ON SITE

Ports, shipyards and vessels are high risk environments, especially during construction phase or during large maintenance works. This is mainly due to several parallel work activities going on as well as the nature of the works, such as welding and painting which can introduce high risks *e.g.* fire on board. Some ports also have an increased risk level due to the goods being transported, *e.g.* chemicals. This code of practice (COP) has been developed to help MARINE employees and sub contractors to be prepared in case of emergency at site.

SCOPE

These requirements apply to all Marine and Cranes employees and contractors working on site or visiting site without guide. All personnel must have essential contact information with them and follow the site's/customer's protocols in case of emergency.

HAZARDS AND RISKS

Working on board a ship or on a crane (only one way of exit and no light) presents an increased risk in respect of evacuation if *e.g.* a fire breaks out or there is a chemical threat. By knowing how to act if such incidents should occur before they do, one will be able to handle the situation better.

OPERATIONAL CONTROL

All MARINE personnel that are going to work on site or visit it needs to know the Emergency Contact Plan and the Escape Plan for the site. Personnel working on site also need to know the project's Risk Assessment, Project Safety Plan and site's/customer's Emergency Preparedness plan

Escape Plan

This plan should be a map of the area marked with the locations of muster points, firefighting equipment, first aid stations and escape routes. It should also describe how the site warns people if an emergency occurs, like bells ringing when there is a fire. This map can also show other useful site locations, like toilet and canteen.

Emergency Preparedness Plan

The site or the customer should provide MARINE personnel an Emergency Preparedness Plan prior to work start, or its content can be covered during site specific safety introduction. If this is not provided, MARINE PM needs to get a hold of the information and inform the personnel that are going to work there about what to do in case of critical situations. The Emergency Preparedness Plan should inform about what to do and who to alert in case of emergency on site, *e.g.*:

- Fatal or serious injuries
- Fire evacuation on board a vessel or crane
- Evacuation at sea

- Chemical spillage to the environment or on board
- Forces of nature like hurricane, tornado and earthquake
- Emergency related to work in confined space
- Emergencies that can happen when travelling to high risk countries

A good emergency plan should also include key procedures, such as:

- STOP all the work if a emergency occurs
- Shut off the valves and switches close to you if you can
- WALK (DO NOT RUN) to the nearest designated muster point
- Stay there for head account
- Only go back to work when ALL CLEAR SIGNAL is given
- Always bring with you a flashlight when on board for a safe evacuation

Risk Assessment

The Emergency Contact Plan should cover operations/activities that have high scores in the project EHS Risk Assessment.

Project Safety Plan

The Project Safety Plan is the main safety document for the project, so this document should refer to all the above mentioned documents and link to them.

COMMUNICATION

All MARINE persons who are likely to have to travel to customers' sites shall be briefed on the contents of this COP.

PERSONAL PROTECTIVE EQUIPMENT

This code of practice in respect of the selection, provision and maintenance of personal protective equipment (PPE) has been developed for Marine and Cranes activities both for operations within the ports and, shipyards but also whilst at sea during commissioning.

SCOPE

These requirements in respect of PPE apply to all Marine and Cranes employees and/ or contractors working on customers' sites. Requirements for respiratory protection equipment (RPE) are covered in COP 29.

If the local regulatory or customer requirements impose a higher standard, then the higher standard shall be followed. In such cases this shall be documented.

HAZARD & RISKS

Work instructions shall instruct in safe working methods and requirements for PPE. PPE shall be considered as the last resort protection measure, *i.e.* other safety measures should always be considered prior PPE. In most projects MARINE is not fully in control of the work site, since the shipyard/ vessel owner/ vessel operator carry the main responsibility.

There may also be several contractors working in the same area as MARINE employees. Thus the usage of PPE should always be considered based on the actual situation at site.

OPERATIONAL CONTROLS

Selection & provision of general PPE

All PPE that is to be provided for Marine and Cranes personnel including any possible sub contractors must be fit for purpose and comply with the relevant EN or ANSI standard or their equivalent. The following represents an indicative list of PPE or other safety equipment that shall be provided for all persons who have to visit and work at customers' sites. Personal issue

- Safety helmet or safety helmet equipped with visor
- Ear protection (personal or disposable)
- Respiratory protection-disposable dust mask or reusable half mask with P3 filters
- Eye protection (glasses, goggles or face shields)
- Overall or cover suit – flame resistant
- Protective footwear-boots
- Protective Gloves
- Flashlight with spare batteries
- MARINE lock out/tag out equipment

Additional safety equipment to be available at site

- Danger signs and hazard tape
- Portable earthing devices
- Insulating mats
- HV indicator rods
- Walkie talkie

Care and Maintenance

All non disposable PPE shall be maintained by each employee so that it is kept in good condition for use. In certain cases such as hearing protectors then they need to be checked periodically to ensure that the seals are being kept in good condition. Similar comments apply to eye protection, safety footwear and head protection. In the case of disposable PPE such as ear plugs then the supervisor on site shall ensure that there are sufficient supplies kept and available on site.

Storage

All MARINE staff that is required to travel to site shall be provided with a suitable bag in which to store all the required PPE to ensure that it is kept in good order and is prevented from being damaged in transit to site.

LONE WORKING

There is no general prohibition of lone working but wherever it is practical lone working should be avoided. This code of practice has been developed as general guidance for those occasions when it may be necessary.

SCOPE

This COP applies to all Marine and Cranes employees and contractors, or work sites which have a nominated MARINE Site manager or responsible person. If the local regulatory or customer requirements impose a higher standard, then the higher standard shall be followed. In such cases this shall be documented.

HAZARDS AND RISKS

There is no additional hazard that arises from lone working but the risk of serious injury or illness arises from the fact that the worker is on his own and he cannot call for assistance and as a consequence the outcome from any incident will be more severe. This might include a work related incident or a medical condition.

OPERATIONAL CONTROLS

As in all cases risk assessment shall have been undertaken prior to any work starting and preferably during the planning stage when the possible lone working situation will have been identified. The assessment shall have identified the hazards associated with the proposed work activity and specified the control measures required in the normal way as per the hierarchy of control. Where lone working cannot be avoided then proper consideration shall be given to risk reduction measures. They shall include consideration of the following:

- Avoid the need for lone working if at all possible. If necessary postpone the job.
- For high risk activities such as live electrical work, work in confined spaces, work at height etc lone working is prohibited. Live electrical work means working either on or in close proximity to energized conductors.
- *On no account is lone working permitted when working on or near to electrical installations where there is a foreseeable risk of electrocution or in confined spaces, as there shall be suitable arrangements in place to provide emergency support and first aid, or to arrange for its provision.*

- Ensure that the lone worker is competent and understands the limits of the work activities.
- Ensure that the MARINE employee registers his presence on site with the security office on site and that he informs them about his work location.
- Reduce the risk by limiting the activities to low risk ones such as monitoring duties only and no operation of plant, equipment or machinery, working at height, operating vehicles etc.
- In cases where an engineer may be required to work on electrical equipment such as a drive for a large crane then that person shall be accompanied until the seven steps have been properly applied and the equipment is proved dead and the isolator has been locked off.
- In a similar way the equipment cannot be re-energized until a second person is present.
- Provide a suitable means of communication so that the lone worker can alert a contact person or someone else on site in the event that there is something of concern, or in the case of an emergency. The person need not necessarily be an MARINE employee.
- Provide the lone worker with some form of alarm, manual or automatic which operates when there is a lack of activity.
- Actively contact the lone worker to check on his status throughout the shift, or ensure that there is a regular call in schedule.
- In cases where the lone working situation arises unplanned then the lone worker shall contact his supervisor to agree the precautionary measures that need to be applied. He should also contact the customers' representative on site.
- Ensure that the lone worker has a "travelling" first aid kit available.

ELECTRICAL SAFETY

Working on electrical systems at voltages above 50V will present serious hazards to persons both in terms of electrocution, but also in respect of the effects of arc flash which can occur on systems below 1000V as well as >1000V. This code of practice describes the basic minimum requirements for electricians/engineers of Marine and Cranes working on electrical systems in a site environment. Further more detailed information is provided in the required electrical safety trainings and in the Group instructions.

HAZARDS AND RISKS

Work undertaken on electrical equipment will present a number of basic hazards. This can include electric shock which if the person is subject to a voltage across his body can result in serious injury and often death depending on the voltage, and the current and time involved. This can occur at currents of 30mA. There is also the hazard of arc flash which occurs if an accidental short circuit occurs when a conductive object gets too close to a high amp (power) current source, or by equipment failure. It can often result in the release of very high energy levels over a very short time period releasing large amounts of heat with the conductors becoming molten, which are then vaporised. This can result in severe burns to the uncovered parts of body. It is particularly relevant at voltages below 1kV owing to the compact nature of the equipment and therefore the small clearances but can also occur at voltages above 1kV. Other hazardous effects include exposure to ultraviolet radiation, effects of arc blast (pressure wave) and possible fire.

SCOPE

This COP applies to all Marine and Cranes employees and contractors working on customers' sites. If the local regulatory or customer requirements impose a higher standard, then the higher standard shall be followed. In such cases this shall be documented.

Application of the 7 Steps

The 7 steps shall be applied to all electrical work carried on Marine & Cranes work sites.

- STEP 1 - CLEARLY IDENTIFY THE WORK LOCATION
  - A caution/danger notice or label shall be posted on the isolator and also on the conductor(s) to indicate that it is being worked on.
  - It is important to be able to distinguish between the equipment that is dead and other equipment that may be live.
  - Make sure that the possible live equipment are in good order
- STEP 2 - DISCONNECT AND SECURE AGAINST RECONNECTION
  - Disconnect equipment from supply and ensure that the isolator is locked in the OFF position. If a number of people are working on the system then a multiple locking hasp and warning notice shall be used so that each person can apply his personal lock. This is often referred to as "Lock out - tag out".
  - If isolation has been achieved by removing a fuse then ensure that the fuse is removed and held by the supervisor and that the fuse cabinet is locked and the key retained by the supervisor
- STEP 3 - PROTECT AGAINST OTHER LIVE PARTS
  - Where there are adjacent conductors that may be live then additional screening measures will be required to ensure that no contact is possible.
  - In the case of HV work (>1kv) safe distances shall be established by suitable barriers, to prevent the risk of contact.

- STEP 4 - TAKE SPECIAL PRECAUTIONS WHEN CLOSE TO BARE CONDUCTORS
  - Wear relevant personal protective equipment including flameproof overalls, gloves rated at the correct voltage and face & neck protection.
- STEP 5 - CONFIRM THAT THE INSTALLATION IS DEAD
  - The circuits to be worked on shall be proved as dead by testing, at the point of work. The testing device itself shall be tested immediately before and after testing;
- STEP 6 - CARRY OUT EARTHING AND SHORT CIRCUITING
  - In addition to the above apply properly designed earthing devices to the conductors being worked on.
  - On transformers it is important that both sides of the transformer have been isolated and earthed.
- STEP 7 - ISSUE THE PERMIT TO WORK
  - The use of a written permit to work linked to the lock off device shall be drawn up and issued by the supervisor who shall witness the testing of the conductors to ensure that they are dead and that they cannot be made live once work has started, *e.g.* no possibility of back feed or induced voltages.
  - The permit to work shall be issued to the nominated person in charge of the work.
  - Supervisor shall check on the work periodically to ensure that there is total compliance with the safety requirements;

Arc flash protection

All electricians in Marine and Cranes shall be provided and shall wear the personal protective equipment specified by MARINE where there is a potential for arc flash. This shall include the provision of any or all of the following:

- suitable head protection EN 50365
- eye/face/neck protection EN 166
- suitable voltage rated gloves EN 60903 or ASTM equivalent
- whole body clothing EN ISO 11612 (flame proof)
- safety footwear EN531
- hearing protection
- voltage rated tools VDE EN 60900.

TESTING & COMMISSIONING EQUIPMENT

Incorrect usage or use of defective/inadequate testing and commissioning equipment can lead to serious injuries. This Code of Practice has been prepared to provide basic guidance on the safe use of devices used in testing and commissioning of LV equipment.

SCOPE

These requirements apply to all Marine and Cranes employees and contractors when working on site. If the local regulatory or customer requirements impose a higher standard, then the higher standard shall be followed. In such cases this shall be documented.

HAZARDS AND RISKS

An insulation tester, also known as megometer and megger, is a portable device used to test wire insulation. Wrong usage of an insulation tester can lead to electrocution.

The shock should not be fatal for people unless they have some heart issues, but it can lead to dangerous situations like losing consciousness while standing on scaffolds. A multimeter, also known as multitester and VOM (Volt-Ohm meter), is used to measure voltage, current and resistance. Incorrect usage of an unsafe multimeter can in worst case lead it to explode. Multimeters are rated into categories based on their intended application. The four categories are:

- Category I: Used where equipment is not directly connected to the mains
- Category II: Used on single phase mains final sub-circuits
- Category III: Used on permanently installed loads such as distribution panels, motors and 3 phase appliance outlets
- Category IV: Used on locations where fault current levels can be very high, such as supply service entrances, main panels, supply meters and primary over-voltage protection equipment

OPERATIONAL CONTROLS

Insulation tester

- Completely discharge, drain the circuit and check for dead before connecting an insulation tester to it
- Disconnect the device to be tested from other electrical circuitry before testing
- Cables and capacitors can store charge for some time after test, and they can be charged with dangerous amounts of energy from a low output insulation tester. After testing, discharge the equipment/the measured circuit by using a cord between the circuit and ground, while wearing voltage rated gloves.
- Never touch the test leads
- Ensure that the insulation tester is within valid calibration date
- To avoid damage on equipment, do not use an insulation tester to

test a low voltage-resistant device, unless the insulator has low resistance range. Insulators should mainly be used to test insulation measurements or conductor capability.

Multimeter

Before travelling to site:

- Test the multimeter at regular intervals and check that the calibration is valid
- If possible, choose a multimeter that has protection against overload on the ohm function and with protected entry connections
- Only use test cords with 90 degrees connections to the multimeter, protection against physical contact and different colors
- Use a multimeter with fused leads (500mA) and probes with minimum exposed metal on the tip to avoid short circuit across terminals or to earth during testing
- If possible, ensure that the multimeter complies to accepted safety standards (UL 3111, EN 61010, IEC 1010 (replaces IEC 348))

Before using the multimeter:

- Be sure that the multimeter is fully functional, test the functions before usage
- Check the test cords for physical damages
- Choose the right setting and range for the specific measurement

When measuring:

- Use the three-point test method, especially when checking to see if a circuit is dead. First, test a known live circuit. Second, test the target circuit. Third, test the known circuit again. This verifies that your meter works properly before and after the measurement. If possible, use a dedicated instrument (preferably with an attached "safe" proving device) to test for dead circuits, since a lot of things can go wrong with a multimeter.
- Hook on the ground clip first, and then make contact with the hot lead. Remove the hot lead first, the ground lead last.
- Hang or rest the meter if possible. Try to avoid holding it in your hands, to minimize personal exposure to the effects of transients.
- Keep one hand in your pocket when possible. This reduces the chance of a closed circuit across your chest area and through your heart.

Tools

Use tools intended for electricians (*e.g.* fully insulated screwdrivers except for the tip). The maximum voltage limit should be stated on the tool.

SAFE USE OF LADDERS

Working at height including use of ladders is one of the major areas for accidents and incidents. This code of practice provides general guidance on the safe use of ladders and stepladders on work sites.

SCOPE

This COP applies to all Marine and Cranes employees and contractors. If the local regulatory or customer requirements impose a higher standard, then the higher standard shall be followed. In such cases this shall be documented.

HAZARDS AND RISKS

The main hazard associated with working from a ladder is one of falling as a result of a number of possible causes.

Typical causes of incidents involving ladders include:

- Ladders are defective because they have not been inspected nor maintained;
- "Home made" or makeshift equipment has been used;
- Placing of ladders on uneven surfaces or where they could slip easily;
- Ladders are in use close to exposed live electrical conductors including overhead lines;
- Persons working from ladders overreach causing them to fall;
- Use of tools that require a high degree of leverage causing a person to over balance and then fall.

OPERATIONAL CONTROLS

Risk avoidance/elimination

As with all health and safety requirements the hierarchy of control should be applied where risk avoidance should be the first consideration. A significant number of serious incidents occur each year involving ladders and hence they should not generally be used as a working place except for simple work of short duration *e.g.* 30 mins.

Significant work of longer duration will require other measures and include the need to consider other methods for working at height *e.g.* fixed platforms, mobile scaffolds or mobile elevated working platforms (MEWP).

Risk reduction and control

Ladder pre-use inspection

- Always check a ladder before using it.
- Check all ladders to see that steps or rungs are tight and secure.

- Be sure that all hardware and fittings are properly and securely attached.
- Test movable parts to see that they operate without binding or without too much free play.
- Inspect metal and fibreglass ladders for bends and breaks.
- Never use a damaged ladder. Tag it "Defective" and report it to your supervisor (to be removed).
- Ladders should also have an ID.

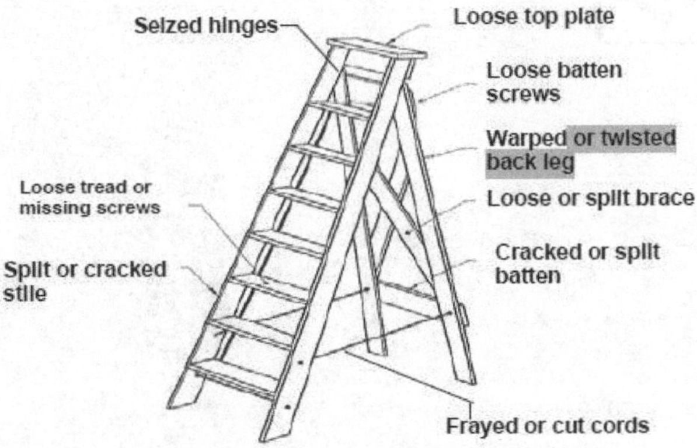

If the ladder is fitted with metal spreader arms check that they lock into place when fully spread.

Fig. Inspection items on ladders

Ladder set up

- Place ladder firmly and evenly on the ground with a slope of 4:1. Make sure the ladder is sitting straight and secure before climbing it. If one foot sits in a low spot, build up the surface with firm material.
- Do not try to make a ladder reach farther by setting it on boxes, barrels, bricks or other unstable bases.
- Do not allow ladders to lean sideways. Level them before using.
- Brace the foot of the ladder with stakes or place stout boards against the feet if there is any danger of slipping or get another person to foot it.
- Never set up or use a ladder in a high wind, especially a lightweight metal or fibreglass type. Wait until the air is calm enough to ensure safety.
- Never set up a ladder in front of a door unless the door is locked or a guard is posted.

- Do not use ladders on ice or snow unless absolutely necessary. If they must be used on ice or snow, use spike or spur-type safety shoes on the ladder feet and be sure they are gripping properly before climbing.
- Use safety shoes on ladder feet whenever there is any possibility of slipping.

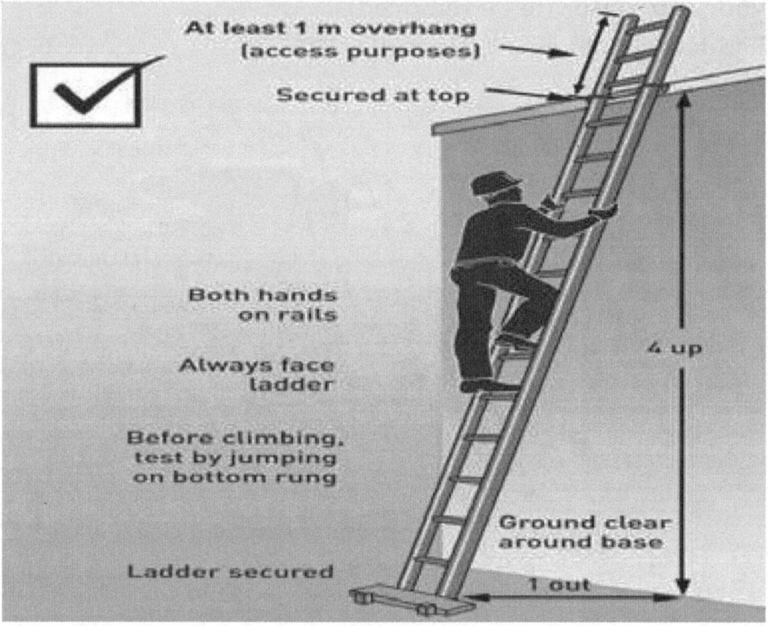

Fig. Correct Ladder Set Up

Ladder Climbing & Standing-key points

- Keep the steps and rungs of ladders free of grease, oil, wet paint, mud, snow, ice, paper and other slippery materials. Also clean such debris off your shoes before climbing a ladder.
- Always face a ladder when climbing up or down. Use both hands and maintain a secure grip on the rails or rungs, always three points of contact.
- Never carry heavy or bulky loads up a ladder.
- Climb and stand on a ladder with your feet in the centre of the steps or rungs.
- Do not overreach from a ladder, or lean too far to one side. Overreaching is common cause of falls from ladders. A good rule is to always keep your belt buckle inside the rails of a ladder. Work as far as you can reach comfortably and safely, and then move the ladder to a new position.

- Never climb onto a ladder from the side, from above the top or from one ladder to another.
- Do not use rungs as hand grabs.

Fig. Examples of ladder use

Correct use

- Never use metal ladders around exposed electrical wiring. Metal ladders should be marked with tags or stickers reading "CAUTION-Do Not Use around Electrical Equipment" or similar wording.
- When using a ladder where there is traffic, erect warning signs or barricades to guide traffic away from the foot of the ladder. If this is not possible, have someone hold and guard the bottom of the ladder.
- Do not try to move a ladder while you are on it by rocking, jogging or pushing it away from a supporting wall.
- Do not leave tools or materials on top of ladders.
- Never push or pull anything sideways while on a ladder. This puts a side load on the ladder and can cause it to tip out from under you.
- Allow only one person at a time on a ladder unless the ladder is specifically designed for two people.
- Never use a ladder as a horizontal platform, plank, scaffold or material hoist.
- Never use a ladder on a scaffold platform. If you need to reach higher, the scaffold should be higher.

Fig. Correct Use of Ladders

WORKING HEIGHT

This code of practice (CoP) has been developed to provide practical information to Marine and Cranes personnel and contractors on the hazards and risks involved when working at height. Special attention is made on works performed under a ship, but the same principles apply also to other environments where working at height features. Definition for work at height in this context is working in heights more than 1.8 meters (6 ft).

SCOPE

This COP applies to all Marine and Cranes employees and/or contractors working on customers' sites. If the local regulatory or customer requirements impose a higher standard, then the higher standard shall be followed. In such cases this shall be documented.

HAZARDS AND RISKS

The hazards and risks associated with working at height under a ship are the same as other environments where working at height features. The main difference is that the equipment used to provide a safe working platform is usually owned by the shipyard and can very often be in a poor state of repair. This is particularly the case in countries which do not have a high standard of EH&S traditionally. Hazards therefore include:

- Falling from height;
- Falling objects

OPERATIONAL CONTROLS

Risk avoidance

As with health and safety risks the hierarchy of control should be properly applied and risk avoidance is the best and most preferred option. Therefore consideration should be given at the planning stage as to what work(if any) can be undertaken without having to work at height. If risk avoidance is not a practicable option then consideration will then need to be given to risk reduction and risk control possibilities.

Risk Reduction & control

When considering options for working at height there can be a number of possibilities. These include the use of:

- Custom made equipment
- Fixed scaffolding
- Mobile scaffolding
- Use of mobile elevated working platforms
- Use of fall arrest equipment.

Custom made working platform

A custom made working platform designed specifically to provide a safe means of access to the Azipod units is a preferred option in that it avoids the necessity of having to erect the working platform each time it is required.

Fig. Custom Made Equipment

It does require to be maintained to ensure that it has not become damaged etc and is no longer unsafe to use.

Fixed scaffolding

The main duty is to provide a safe place for persons to work from so that they cannot fall from height whilst carrying out their work. A fixed platform/ scaffolding can provide an effective means by which this can be achieved but it has to be erected by a person who is competent, it has to be in a good state of repair and has to be complete.

(a) Sound footing (b) Vertical upright +bracing
(c) Secured by ties (d) platform, toe boards/rails

Fig. (a) and (b) Fixed Scaffold in a dangerous state

Key requirements are that:

- Scaffolding shall erected by a competent person'
- Components shall be in good condition particularly load bearing ones
- Vertical members shall be upright and have a secure footplate
- Lateral bracing to be provided
- Scaffolding to be tied or otherwise secured to the structure-*e.g.* ship.
- Adequate platforms in good condition to be provided and which are complete.
- Guard rails to be provided at 950 and 470mm
- Safe means of access to be provided and secured.
- If for any reason a scaffold is considered to be unsafe then it should be tagged as such.
- It shall be inspected before first use and thereafter if modified etc.

Mobile scaffolding

Mobile scaffolding often have same faults or failings as other types of scaffolding.

These include:

- Component parts become damaged but are still used.
- They are poorly erected with no adequate means of access,
- Poor platforms with limited numbers of boards in place or poor quality boards;
- No toe boards and no guard rails or poor guard rails;
- Wheels not locked in position;
- If sheeted then wind loading not considered adequately
- Not inspected nor maintained when mobile tower is moved or modified.

Fig. (a)-(c) Damaged and Dangerous Mobile Scaffolding

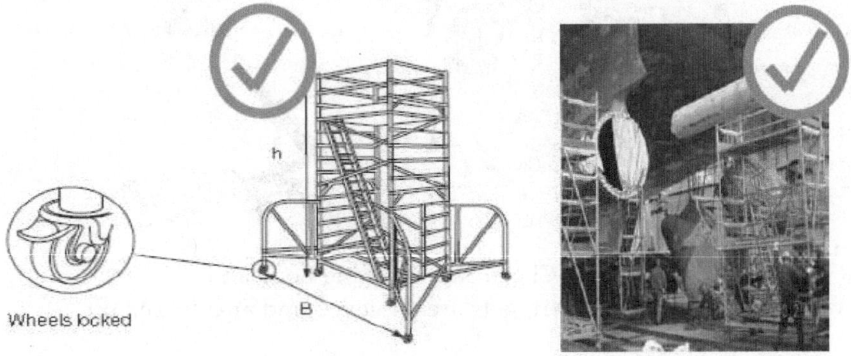

Max ratio h:b=3:1

Fig. Correct Set up-Mobile/Modular Scaffold

Mobile elevated working platforms

Mobile elevated working platforms (MEWP) provide a safe means of getting access to difficult areas beneath a ship. They can vary in size from small scissor lift type machines to larger self propelled ones. Basic precautions for MEWP are:

- Ensure surface is even (flat) and stable and able to withstand the weight the machine. Steel plates may be required if the bottom of the dry dock is uneven or unstable for any reason.
- In areas where other persons may be working, or the work may be interfacing with the customers employees then the area shall be barricaded to keep unauthorized persons out of the area.
- Check the safe working load.
- All operators of MEWP equipment shall wear suitable PPE as follows:
  - A fall arrest harness attached to a lanyard with a shock absorber (ISO 10333/ANSI Z359.13) shall be worn at all times when in the basket of the elevated work platform. The lanyard shall be attached to the designated anchor point in the basket and attached to the personal harness at all times.
  - Safety helmet
  - High visibility vest or equivalent
  - Safety boots.

Whenever a person is wearing a harness, an elevated work rescue plan shall be in place and a person in position with the knowledge to switch the elevated work platform to ground controls and lower the elevated work platform to the ground

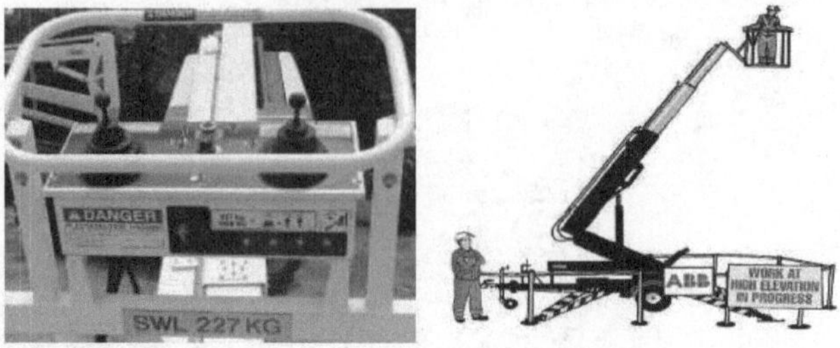

Fig. (a) Check Safe Working Load and
(b) Ensure that outriggers are deployed and area fenced off

Fall arrest

Fall arrest equipment should not normally be required if a safe working platform has been provided either by means of a fixed platform or by means of a mobile or modular platform. If fall arrest equipment (fall arrest harness attached to a lanyard with a shock absorber) needs to be used, it shall comply with ISO 10333 and persons shall be instructed in its use. When fall arrest equipment is used, the safe method of working shall also consider how the rescue will be done in case it becomes necessary.

Lighting

A dry dock will generally not have lighting installed and hence it is important that any project considers the need for adequate lighting. Poor lighting will lead to persons not being able to see the working platform and the safe means of access and there will be much shadow.

It is important that lighting is provided for the work place itself and the means of access to it. Lighting supplies should be protected with Ground Fault Circuit Interrupters or otherwise safe against cable damages etc.

COMPRESSED GASES IN A SHIP

Compressed gases supplied in cylinders are in common use throughout industry and in particular in the marine sector. In a shipyard environment compressed gases are more likely to be supplied via manifolds and multiple hoses.

Their use on board ship however can bring a number of significant hazards and this code of practice has been developed to provide information about the safe, handling and use of these materials when working on board ship.

SCOPE

This COP applies to all Marine and Cranes employees and contractors working on customers' sites. If the local regulatory or customer requirements impose a higher standard, then the higher standard shall be followed. In such cases this shall be documented.

HAZARDS AND RISKS

General

Cylinders containing compressed gases can present a range of hazards which include:

- Effects of blast of gas under pressure
- Direct effects of the contents *e.g.* toxic, flammable, corrosive etc.
- Fire or explosion in respect of flammable gases, *e.g.* propane
- Manual handling injuries from the movement of cylinders or from cylinders falling
- Leaking of gas resulting in build up of concentration of gas. *e.g.* oxygen enriched atmosphere in confined spaces

Confined space working

Working within a ship however can result in the above hazards increasing in significance because of the environment. The main point is that a ship below decks is often a confined space and hence the use of gases in this environment will increase the overall hazard. Most gases are also heavier than air hence they will collect in compartments or other similar places, such as the Azipod unit. For work inside Azipod unit a separate rigorous risk assessment must be carried out including checking on the atmosphere in the confined space.

Fire & explosion

Chemical properties of certain gases bring with them significant fire and explosion risks. Oxygen enrichment is a hazard where the concentration of oxygen in air increases with the real possibility of a fire and explosion. This is a particular hazard on board ship as if oxygen is allowed to collect then the fire characteristics of materials will change dramatically leading to a serious fire. The increase of oxygen in air need only be 1% before this becomes a reality. Propane and butane are two liquefied petroleum gases (LPG) and have wide use. Propane is generally used for flame cutting and burning and butane tends to be used for heating purposes.

If not used carefully then serious incidents can result ranging from small burns or fires to serious explosions. The vapours are also heavier than air and will collect therefore in ships compartments, ducts, bulkheads etc. Both propane and butane when vaporized will produce a large volume of gas at normal temperature and pressure, 230 times for butane and 270 times for propane. Example: 1 litre of liquid propane, when vaporised, will produce

270 litres of gas. The gas mixed with air at its lowest explosive limit of 2% will produce 13,500 litres of flammable mixture which if ignited will result in a major explosion.

OPERATIONAL CONTROLS

General

The following represents best practice for storage, transportation and handling of compressed gases in cylinders:

- Compressed gases must have labels to identify their contents;
- Gas cylinders, especially the valves are protected from weather, *e.g.* rain;
- Gas cylinder storage area shall have good natural ventilation and cylinders should be stored in upright; position, on a flat surface and secured against falling over;
- Smoking shall be prohibited in the storage area and firefighting equipment shall be readily accessible
- Fuel gases shall stored separately from oxygen (more than 1m apart)
- Cylinders should be lifted using suitable cradles and slings when using a hoist or crane;
- Pipe work connection checks are carried out to ensure that connections are of good quality with crimped fittings
- Checks are carried out after use to ensure that valves always closed after use
- Hoses are checked periodically for wear and damage
- Flashback arrestors are used on cylinders containing fuel gases.

Labelling

There is no internationally recognized standard for color coding of gases but within Europe there is an EN standard (EN1089). There is also UN standard for hazard symbols. Gases are color coded according to the risk diamond on the main label.

- Company name
- Address of the company
- Risk & safety phrases
- Hazard symbols(UN classification)
- EEC label
- Revision number
- EEC number if applicable
- Product name

- UN identification number and proper shipping name
- Additional information

ENTRY INTO AZIPOD UNIT

This code of practice (COP) has been developed to provide practical information to staff within Marine and Cranes and contractors on the hazards and risks involved while entering an Azipod, related confined spaces and the required controls. This COP does not provide any permit to access Azipod and is not all-inclusive. Persons entering the Azipod confined spaces must be authorized and trained to do so and required controls must be in place. For further details contact your safety advisor.

Examples of possible confined spaces onboard:

- Azipod & Cooling Air Unit & Slip ring unit
- Double bottom & compartments in ships
- Storage tanks
- Rooms or areas where there is no ventilation or poor ventilation *e.g.* cable ducts

How to Identify a Confined Space?

- Limited openings for entry & exit
- Unfavourable natural ventilation
- Not designed for continuous worker occupancy

This COP applies to all Marine and Cranes employees and contractors working on customers' sites. If the local regulatory or customer requirements impose a higher standard, then the higher standard shall be followed. In such cases this shall be documented..

HAZARDS AND RISKS

While working inside Azipod unit, the employees could be exposed to the following hazards (but not limited to):

- Atmospheric hazards
- Hazards due to moving mechanical parts
- Electrical live parts
- Hazards due to vertical entries
- Potential oil contamination possibly causing skin irritation
- Thermal hazards (*e.g.* high temperature of propulsion motor)
- Inconveniences due to confined space.

OPERATIONAL CONTROLS

Risk Avoidance

As with health and safety risks the hierarchy of control should be properly applied and risk avoidance is the best and most preferred option. Therefore if it is possible to perform the work without entering the Azipod Unit, this working method should be preferred.

Risk Reduction & Control

The following actions are required to ensure protection of workers from the hazards that can be present in confined spaces.

Entry for Rescue

Entry for rescue requires certified Azipod space entry training, ensuring persons capability to use the rescue equipment and correct rescue methods. In case the rescued people is unconscious or confused *i.e.* there are any signs that the atmosphere might be endangered, the atmosphere must be secured prior entering for rescue.

7

Shielding the Effectives

PROTECTING THE OPERATIVE

BASIC WELFARE PROVISIONS

Those consulted as part of the project considered that there was often insufficient consideration given to basic welfare provisions. Coastal construction sites are often situated in remote locations where there are little existing welfare provisions. While records are maintained on lost time due to safety incidents, there is little or no record of lost time due to sickness and general poor health or absenteeism caused by unsociable hours/poor working conditions.

It is essential therefore to consider:

- How basic welfare and cleanliness provisions will be delivered to the operatives in a safe manner
- How the works will protect against disease and contamination risks (*e.g.* contaminated dredged material, Weils disease, etc.)
- How tidal working/unsociable hours might affect the operatives. health and the subsequent management of health and safety of the site. Long shifts and tidal working rotating around the clock, often mean that operators are obliged to work in poor weather and where it may be difficult for them to visit the mess hut
- How the safety of any lone-worker operations in remote areas will be delivered
- How fatigue and stress to the workers will be avoided and mitigation measures managed.

The above are significant issues in the coastal and maritime sector and failure to provide and consider basic welfare provisions will cost businesses money and delay.

PARTICULAR PROVISIONS

The following section is based on guidance developed by the MCA but extended following consultation with the Project Partners.

Good practice: Particular provisions for working over water

Platforms, gangways, etc.

Shore-supplied platforms and gangways must comply with the requirements of the Construction (Health, Safety and Welfare) Regulations 1996. At all edges from which a person might fall into water, platforms, guardrails, barriers, etc. are required. Ship-supplied platforms and gangways must comply with Merchant Shipping legislation. Warning notices regarding the location of the water.s edge should be erected at all unguarded edges and boundaries near water and set so that operatives approaching the water.s edge can easily see them. Where platforms or gangways are erected above tidal water, decking boards should be secured (clipped or wired down) so that rising water or high winds cannot dislodge them. The provision of additional handholds is always advisable as a precaution in the event of storms.

Safety nets

It is often better to use physical barriers rather than nets. However, where used these should be properly secured and slung sufficiently far above high water level for anyone caught in them to remain clear and so that free access of rescue craft is always possible. Manufacturers should be consulted on what type of net best suits the kind of work to be carried out and the prevailing conditions.

Harnesses

Harnesses should only be used where no other protective measure can be provided as they can pose an additional tripping hazard and potential fatal suspension trauma. Barges, pontoons, etc. used as working platforms, must be fit for purpose, properly constructed and sufficiently stable to avoid tipping. Attention must be paid to good anchorage and ballasting; point loads near the edge should be avoided; due account should be taken of the variation of load at the different radii of crane jibs for the use of lifting appliances on barges and pontoons.

Lifting equipment

Handling of materials and equipment over water often requires the use of floating plant. Consideration must be given to the operational limits of such plant, in terms of wave and current forces as well as operable wind speeds. Particular attention is required when transferring materials to/from a floating platform to a fixed platform because of relative movements between the two.

Site tidiness

Site tidiness is key in minimising tripping hazards. Tools, ropes and other materials not in use should be stored away; rubbish should be cleared up promptly. Materials awaiting use should be stacked compactly and restrained, particularly on pontoons and boats. Mooring lines should be clearly marked and protected. Slippery surfaces are dangerous and should be treated immediately. Seaweed, sea-slime and bird droppings should be cleaned off. Oily or greasy surfaces should be gritted. Icy or frosty surfaces should be treated with industrial salt or sand. Drip trays should be sand-filled and set beneath all machinery to prevent the development of oily, slippery surfaces and, especially on pontoons, to minimise fire hazards. The water area near the work should be kept clear of flotsam. Boatmen should be instructed to report its presence and to clear where possible.

First aid equipment

First aid facilities should be available, in the charge of a first aider or appointed person, on pontoons, barges and near all landing places.

Means of access: water transport

The Construction (Health, Safety and Welfare) Regulations, 1996 require the safe transport of any person conveyed by water to or from their place of work. Passenger-carrying craft must be appropriate for the number of persons being transported. Such craft should be marked with the number of persons they are intended to carry and, where appropriate, the limits of operation. Where craft are of sufficient size to carry more than 12 passengers at any one time, they must be surveyed at least once each year by the MCA and a certificate obtained. Boats are only allowed to operate with a stipulated number of passengers according to size and the MCA makes several classifications of craft according to purpose, with a schedule of life-saving and fire-fighting appliances laid down for each. There are currently no specific requirements relating to the competence of the boat operator, although they must be competent to operate the craft being used. A suitable means of communication should be established with the shore during any work activity and in the event of an emergency.

Manning of work boats and similar craft

The.Code of Practice for the Safety of Small Work Boats and Pilot Boats. applicable to the manning of sea-going vessels of less than 24 metres in length, which carry not more than 12 passengers, is available from the MCA. The.Code of Practice for the Safety of Small Work Boats and Pilot Boats. does not apply to vessels operating on inland waters where such waters form part of a river or are completely self-contained such as in a lake or reservoir. In such circumstances further advice should be sought from the river and/or water authority or the local authority regarding any local requirements that might apply.

Emergency planning/preparedness

Particular attention should be paid to effective emergency planning and provision of access to key emergency services such as:

- nearest Hospital
- RNLI
- Harbourmaster
- Coastguard
- Police
- Fire service
- Bomb disposal
- Decompression facility (where diving work is planned).

The safe means of access/egress of emergency services to/from the site, which may involve beach and surf-zone crossings and/or marine access must be clearly identified before the start of operations.

Protective clothing and equipment

Safety helmets

Safety helmets should be worn at all times for operatives in designated work areas, since anyone struck on the head and then falling into water is at special risk.

Footwear

Types with non-slip soles should be worn. Rubber and thigh boots should not be worn as, once filled with water, they act as dead-weight and could drag the wearer under water.

Personal buoyancy equipment

Life-jackets or buoyancy aids must be worn where there is a foreseeable risk of drowning when working on or near water and at all times while working on boats. Risk assessments can be used to identify areas which are safe and where no buoyancy equipment is required (*e.g.* accommodation blocks on a jack-up platform). Equally areas where there is a high risk should reinforce procedures.

A life-jacket is a personal safety device which, when fully inflated (if inflatable), will provide sufficient buoyancy to turn and support even an unconscious person.s, face upwards within five seconds (ten seconds if automatically inflated). The person.s head will be supported with the mouth and nose well clear of the water. Some people are reluctant to wear life-jackets as they find them bulky and restrictive. However, either an automatically inflatable life-jacket or a type which is inflated by a manual pull-cord should overcome these problems. These are usually compact and allow for a full range of movement. Buoyancy aids are worn to provide extra buoyancy to assist a conscious person to keep afloat. However, they will not turn over an unconscious person from a face down position. A wide range of life-jackets and buoyancy aids is available and they should be marked with the relevant BS EN and selection will depend on such matters as the type of water conditions, the work being undertaken and the protective clothing being worn. Further guidance is given in HSE Agricultural Information Sheet No. 1 (Revised): Personal buoyancy equipment on inland and inshore waters, HSE (1999). Where appropriate, life-jackets should be weld protected.

Rescue equipment

Lifebuoys and rescue lines

The MCA approved lifebuoys, or rescue lines, should be set at intervals along the workings. The lifebouys. lines must be checked to ensure that they are long enough to allow for the state of the tide, height of working place above water, or for the person being carried downstream by a current. They may be constructed of either cork with canvas covering, or of polyurethane foam with a rigid PVC cover. Both types are effective in salt or fresh water. If night work is carried out, standard MCA-type selfignition lights should be fitted.

Various types of rescue lines are available. A typical rescue line incorporates 25 m of line in a canvas bag with a small flotation chamber. The end of this line is held, while the bag is thrown towards the casualty. Lifebuoys or rescue lines should be thrown as near as possible to a person in the water; if any tide is running, they should be thrown on the upstream side. All persons involved in the work who may need to use the lifebouys or rescue lines should be instructed in their use and should be given regular opportunities to practice the action to be taken in the event of an emergency. Daily checks should be made to ensure that lifebuoys and rescue lines are still in their proper place and that no repair work is required as a result of vandalism or other interference.

Grab lines

Grab lines, attached to the working place, or at other places downstream, and long enough to allow for the normal rise and fall in tide, can be supplied to give a person something to grab in an emergency. They should be of a buoyant type with a marker float at the free end. Trailing ends of undue length should be avoided so that there is no risk of boats being fouled. Daily checks should be made to ensure that they are still in position and that their condition is sound.

General welfare provisions in an exposed environment

Give proper consideration to:

- how basic welfare and cleanliness provisions will be delivered to the operatives in a safe manner
- how the works will protect against disease and contamination risks (*e.g.* contaminated dredged material, Weils disease, etc.)
- how tidal working/unsocial hours might affect the operatives. health and the subsequent management of health and safety of the site
- how the safety of any lone-worker operations in remote areas will be delivered
- how fatigue and stress to the workers will be avoided and mitigation measures managed.

Flares and other equipment

Flares and other equipment should be provided as risks are identified through the risk assessment process. It is essential that full training for use of all the equipment is given and that the training process itself does not endanger life.

Inter-vessel transfer

Inter-vessel transfer should be avoided as much as possible, restricting access to controlled environments in port areas where full and proper provisions can be made. It is particularly hazardous and a fatality has been subject to investigation by the Marine Accident Investigation Board and there was recent guidance issued by HSE. That guidance states basket transfers to or from offshore installations are considered a high-risk operation. The notice confirms that the Billy Pugh type of personnel transfer basket does not comply with Regulation 5 of the Lifting Operations and Lifting Equipment Regulations 1998 (LOLER) but it can be used in exceptional circumstances, *i.e.* emergencies or when transfer is essential and it is not practicable to gain access by less hazardous means. Alternative forms of basket are available which duty holders should consider if they offer a safer means of personnel transfer for their needs.

There is significant debate about the best method of inter-vessel transfer. Some comments given by contractors consulted are given below. Some contractors prefer the Billy Pugh or Wor Geordie as it can be used on small modular jack-ups and accessed from a RHIB or small work boat. They consider that the Frog is more difficult to access from small vessels. They also consider that the Billy Pugh is more easily and more quickly accessed on the deck of a tug or multicat meaning that the vessel has to hold position for a shorter period of time. Those contractors also considered that the time taken to sit down and strap into the Frog could be detrimental under certain circumstances. The contractors consider that the key to any transfer is ensuring it is properly managed. Any means of personnel transfer between marine craft must be covered by a complete risk assessment and methodology; personnel must wear stipulated PPE, transfer suits and life-jackets and the transfer must be overseen by a responsible person on the jack-up barge using a crane that has a quick rope speed to move the personnel carrier quickly onto and away from the vessel as required.

OTHER ISSUES FOR CONSIDERATION

Use of harnesses

The use of harnesses when working adjacent to water is perceived by many as the most appropriate mitigation strategy. However, suspension trauma has been investigated in detail by the Port of Liverpools.s Safety Manager and reported in HSE.

Issues associated with the use of harness near the water's edge and suspension trauma

A personal fall protection system comprises at least a body holding device, *i.e.* a harness of some type, a lanyard and a reliable anchor. A well thought-out system will seek to minimise the effects of any potential fall. If a fall does occur, that system will arrest the fall with a limited impact force. At this stage, the harness (and the rest of the components in the system) will have stopped the fall, hopefully without it causing any injury. The fall and the arrest of it are only part of the story, and not necessarily the most dangerous. After the fall and its arrest comes the suspension phase, when the casualty either rescues him or herself, if capable, or awaits rescue by another person or persons. After a fall, the body is likely to be in a state of shock. If the casualty is badly injured or unconscious, there is unlikely to be any movement of the legs and there can be serious consequences.

The orientation of the body and the comfort of the suspended person, determined to a large extent by the design of the harness and the position of its attachment point to the system, also play their part in the outcome. The recent review undertaken for the HSE examined the potential problems of the suspension phase of a fall, how the position of the attachment points on various harnesses plays an important role in the comfort and survival of a casualty, and how a selection of harness standards addresses the issues surrounding suspension.

The review reports on existing literature, gives background information to assist the reader, raises issues for discussion and gives recommendations for further work. This investigation identified that, should a person fall and become suspended, it was essential to quickly (*i.e.* within 3/4 minutes) remove a suspended person from that position otherwise significant trauma or even death can occur. Many frequently look at the use of harnesses as a.last resort. where edge protection is not possible but many do not consider the subsequent recovery to a safe place. Others argue that harnesses can cause tripping hazards. Tidal variations also need to be considered when assessing the length of the harness line as it would be very dangerous to allow someone to fall into the water with a harness on.

The Port of Liverpool.s view that has been expressed to the HSE is that under operational conditions it is safer to allow a person to fall into the water than to have them suspended above it. This has been done on the principle of a simple risk assessment based on the circumstances at Liverpool and the means of rescue available for suspended and immersed casualties.

Emergency lighting

The extent of emergency lighting requirements and the residual risks can be an issue on coastal and maritime projects.

Issues associated with emergency lighting near water's edge

One port, operating at night, used normal port overhead lighting to provide a safe lighted environment. The HSE was concerned that the port did not have an emergency back-up system such that if the lighting failed there would, albeit temporarily, be no lighting adjacent to the water.s edge. The HSE placed an improvement notice on the port operator to force them to provide emergency lighting.

The port operator argued that the staff was well trained and familiar with the environment and such a requirement was inappropriate. After discussions and two night time visits from the factory inspector, the improvement notice was lifted when the port operator agreed to invest in high luminosity paint which does provide a benefit for the untrained eye, but as only appropriately trained people are allowed onto the dolphins it is of limited use. The port operator also undertook a trial using photoluminescent strips but this was ineffective, as there was so much background light in.blackout. conditions that it was impossible to see the strips glowing.

PROTECTING THE DIVER

Diving is a potentially hazardous activity in any coastal environment. However, when a good level of planning and preparation is carried out and strict controls applied to comply with information and procedures set out in the preparatory stage, risks can be mitigated and work can be successfully and safely completed. The following sections draw on the requirements of the Diving at Work Regulations (1997) and good practice principles.

RESPONSIBILITIES

The employer who requests the work must ensure that.

- the appointed contractor is competent to carry out the proposed work and is authorised and on the HSE competency list
- the proposed work site is safe and any known hazards, underwater restrictions or associated operational information is made known to the diving contractor
- other appropriate facilities or persons will be made available to the appointed diving supervisor should an emergency situation necessitate.

The competent diving contractor carrying out the works must ensure that:

- a detailed diving project plan is prepared and all risks fully assessed
- details of the plan are made known to anyone directly or indirectly involved in the diving operation or its support activities
- an adequate number of personnel, diving and essential support are available, that sufficient plant and equipment is available and adequate arrangements for first aid or emergency medical support are in place
- the immediate dive site is safe to conduct diving operations and adequate arrangements for the access and recovery of the diver even in an emergency are in place
- accurate records are maintained and all other appropriate regulations are adhered to.

The appointed diving supervisor, in addition to being confirmed in writing and properly qualified to carry out the task, must ensure that:

- everyone involved in or required to support the diving operation is fully briefed and aware of the diving plan and any emergency arrangements that are applicable
- all information, local conditions, weather, tide, current or similar used in assessing the risks on the site remain valid during the period of the dive
- adequate records are kept and persons involved in the support of or working close to the dive site are fully appraised of progress as the operation progresses.

TEAM SIZE

The correct selection of the number of divers necessary to complete a specified task goes some way to ensuring the safe conclusion of any diving operation. While Regulation 6 of DWR and the ACOP indicate the minimum team size to be four. supervisor, diver, standby diver and tender. the Diving Project Plan, in addition to detailing the task and associated risks, will address the requirement for diving team personnel and should specify and justify the team size necessary for a particular task. Basic underwater tasks in controlled circumstances can be safely completed using a four-man diving team, however, as both the complexities in terms of tools, equipment use and depth of the dive increase so too does the likely need for additional members of the dive team.

EQUIPMENT

All commercial diving work should be carried using equipment that is maintained as part of a planned maintenance system and suitable for the work in hand. While DWR 1997 does not exclude the use of Self Contained Underwater Breathing Apparatus (SCUBA), the need to provide communications between the diver and the surface as well the requirements for an independent air supply reserve tend to reduce the benefit of SCUBA and actively encourage the diving contractor to adopt a Surface Supplied Diving Equipment (SSDE) spread. In addition to the benefits of having a larger reserve of air provided from the surface, the working diver using modern lightweight air helmets can be afforded a high degree of head protection while working in the water.

DIVING PROJECT PLAN (DPP)

Consistent with other H&S legislation, the DPP is intended to.. so far as reasonably practical. identify the hazards and risks existing on or at a site and detail how these will be controlled. In addition, the DPP will look specifically at the range of support equipment that will be necessary to cover all foreseeable eventualities on the dive site. The DPP should make clear reference to the working methods that will be used and detail the control measures that will be implemented for dealing with water flows, underwater currents, low visibility, weather conditions, vessel movements; use of tools as well as specifying any special precautions that may be required to ensure liaison between the dive team and other vessels operating in the area. In particular, there is a need to address the method of recovering the diver after completing the work or in an emergency situation when they may be unconscious.

Entry tends to be quite simple but having completed an element of work the ability to climb a vertical ladder may be seriously impaired and some other form of controlled recovery may be required. The DPP will be used to brief the divers and any other persons involved in the diving operation and a copy of the plan should be available to any interested party. Checklists are commonly used as part of a DPP to ensure that all aspects of the duties of the employer, contractor, diving supervisor and divers have been covered.

EMERGENCY ARRANGEMENTS

As with all project planning regimes, ensuring arrangements exist to deal with emergency occurrences is important. In addition to the need to provide a safe work environment, and provide adequate launch and recover facilities, the diving contractor must ensure that adequate arrangements exist for the emergency recompression of divers. Where no in-water decompression is planned and the water depth does not exceed 10 m, arrangements to have a suitable recompression facility accessible within a maximum travelling time of six hours exist.

For dive operations between 10 and 50 m with either no in-water decompression or planned in-water decompression not exceeding 20 minutes, the requirement is reduced into having a suitable recompression facility within a maximum travelling time of two hours. Where diving is planned to require more than 20 minutes in-water decompression, or either of the two requirements outlined above cannot be complied with, a suitable recompression facility will be required on the site. It is worth stressing that for most coastal diving operations, the DPP and risk assessment is likely to demand the need for a recompression facility on site since the sea and weather conditions can change at very short notice thus compromising the ability to achieve either of the two or six hour requirements.

COMPETENCY TRAINING IN COASTAL ENGINEERING

SETTING UP A COMPETENCY MANAGEMENT SYSTEM

Setting up competence management systems and ensuring adequate competence of the operative is a requirement of recent legislation. It is also considered by those consulted to be a key issue in coastal engineering projects. Achieving and maintaining a suitable level of competence for both operatives and supervisory staff requires appropriate training by experienced trainers on a regular and monitored basis.

The training needs may range from on the job training to formal classroom courses and formal qualifications such as NVQs. It is important that training covers not only the site operatives but also the supervisory staff, site agents and engineers, etc., who may require more management-based, rather than practical training, to achieve appropriate competence levels.

EXISTING COMPETENCY MANAGEMENT SYSTEMS

One major problem in the coastal/maritime construction sector is identifying suitable training organisations with the specific experience and expertise in this environment. While there are general civil engineering courses available, these are usually geared towards construction onshore.

In some cases contractors have commissioned their own training courses from such training organisations, but with a particular emphasis on the coastal/maritime construction issues. Smaller specialist companies often link together to share specialist competency training.

CSCS system

The CSCS (Construction Skills Certification Scheme) aims to register every competent construction operative within the UK not currently on a skills registration scheme. Operatives will get an individual registration card which is valid for three or five years. The CSCS card also provides evidence that the holder has undergone health and safety awareness training or testing. Further information can be found on their web site www.cscs.uk.com. The Civil Engineering Contractors Association (CECA) in association with the Construction Industry Training Board (CITB) produces a manual of approved cards. There is currently an exemption for.Marine Construction. but it is anticipated that formal competency schemes will be developed in the future.

IMCA system

The IMCA (International Marine Contractors Association) runs a self regulation competence assurance and assessment scheme to enable contractors to facilitate and improve safety in the offshore industry. This excellent framework enables contractors to assess and demonstrate to others the competence of.safetycritical. personnel. It also includes competency measures for client.s representatives. While it focuses on the offshore industry this is very relevant to coastal and maritime works. Further details including guidance, guidance for assessors, logbooks, etc. can be obtained from www.imca-int.com.

USERS OF THE COASTAL ENVIRONMENT AND PUBLIC SAFETY

The focus of this research is to examine risks associated with the construction, maintenance and demolition process rather than the risks to the users resulting from the permanent works during the use of the structure. However, the research has identified some useful background information which may enable designers to better consider health and safety during the use of the coastal structure and these issues which is presented here.

SETTING UP AND RUNNING A COMPETENCY MANAGEMENT SYSTEM

- Use IMCA competency system to assess competence of staff working on the project.
- Consider setting up a competency system focused around the needs of the coastal sector not just the construction industry. Use staff with recognized experienced in this sector to advise on/formulate the system.
- Ensure competency training is appropriately focused by holding courses targeted on the industry sector. This could be achieved by:

  - Bringing in external trainers but illustrating the training with examples from the coastal sector.
- Use recognised experienced staff to hold internal training courses.
  - Club together with other specialists in the coastal sector to run targeted training.
  - Augmenting standard training with short targeted training.
- Consider formalising the training to NVQ or other recognised accreditation systems.

PUBLIC SAFETY DURING CONSTRUCTION

This section considers health and safety issues associated with the construction, maintenance and demolition process. The coastal zone is, by its very nature, an environment where it is difficult to totally separate the construction process from the public and other users such as fishermen. By their very nature, the high costs associated with coastal defence projects can normally only be justified economically where there are significant assets at risk, with corresponding numbers of people using the area.

Key issues to consider that raise hazards levels when working in close proximity to the public include:

- it is difficult to fence off large areas of coast during construction work
- the public are often attracted to the beach and construction works and may not realise that it is an operational construction site
- dog walkers often will use beaches in the early morning before safety/security guards are monitoring the site
- different users may be expected on coastal/maritime-based developments such as quad bikes, kite flyers, etc.
- marine leisure boat users may be attracted to the construction works and not be aware of the notice to mariners
- in particular, jet skiers and surfers may take a special interest in getting close to floating marine plant
- fishermen often operate outside daylight hours when the site may not be fully manned and they may not be fully aware of the notice to mariners
- establishing when it is it safe to hand back the site to the client and the public.

Potential methods of informing the public about coastal construction sites and protecting the public from entering the site boundary

Experience has shown that the public are less likely to pose a safety risk if they are better informed. The risks to public safety can be mitigated by informing the public through.

- hotlines
- newsletters
- newspaper advertisements/articles
- notice boards
- incorporating safe viewing stations
- school visits
- consultation meetings
- project web sites.

Protecting the site boundaries can be difficult on most coastal engineering projects due to the dynamic environment. However, consideration should be given to:

- security fencing where practicable. This can be targeted to protect active working areas whilst also ensuring non active areas are safe for public use
- security guards/beach marshalls
- buoys to help define the construction zone referred to in the FEPA license and notice to mariners.

PUBLIC SAFETY DURING THE USE OF THE COASTAL STRUCTURE

In 1996 the NRA (now part of the Environment Agency) undertook a review on public safety access to the coastal environment. The overall purpose of that research (Environment Agency, 1996 and 1997) was to assist the authority to define a consistent policy for responding to public safety issues and to provide a system for determining appropriate responses.

While this research has not resulted in any formal policy, it is worth noting that as a result of this work the Agency now has a process of Public Safety Risk Assessment which is mandatory for all new EA projects. The first stage was a diagnosis of the scope of the problem of public safety related to coastal structures, and the review identified potential responses to this problem. The second stage developed a systematic approach to the selection and implementation of hazard mitigation measures and risk management responses to improve public safety.

Research into public safety incidents obtained evidence on 31 incidents from NRA records during the period from the creation of the Authority in 1989 up to June 1994. These are summarised in Figure. The review of past incidents combined an assessment of the views of some of the Authority.s staff and others on perceived hazards, possible outcomes and potential responses. Information on past incidents and current perceptions were combined and analysed, to provide a basis for understanding hazards, their causes and responses. Responses to public safety issues were categorised into following.

- those that modify the hazard
- those that modify public behaviour
- those that modify the outcome of an accident.

At that time the NRA design and operations staff identified the preferred responses to public safety needs:

- control the hazard through safety sensitive design
- prevent access to the hazard
- reduce contact with the hazard
- control the development of the hazard through adequate maintenance.

This is consistent with the approach adopted in Health and Safety legislation. The analysis of current responses by the Authority produced the following observations:

- Responses to public safety hazards were not implemented consistently.
- The use of new construction materials or innovative designs, while being more cost effective than those previously used, may create new safety hazards.

At that time the major components of safety-related spending were ranked in the following order, in terms of their perceived impact on safety:

- use of safety sensitive designs and materials
- maintenance
- patrolling, inspection, monitoring
- signs
- safe routes around hazards.

Stage 2 of the review identified a Safety Management Procedure (SMP) that should be taken up by the Environment Agency. This allowed a structured risk assessment to be carried out using local factors and site specific details. The process allows decisions to be made regarding whether further safety measures are required, and if so what form they should take.

- Step 1: involves the setting up of a Site register ready for use in the assessment process and assembly of site information relevant to safety management.
- Step 2: applies a set of screening tests to determine the appropriate degree of detail of the safety assessment and the urgency or priority of the assessment.
- Step 3: identifies the public safety hazards at the site, taking existing safety measures into account in order to determine how safe the site is by completing the Risk register.
- Step 4: identifies and evaluates responses to public safety risks by

completing the Response register. Assessments of the adequacy of existing safety measures and the merits of alternative or additional responses are made. The most appropriate public safety response for the site is then selected.

8

Electrical Equipment and Installations

INITIAL SURVEY

Surveyors should ensure that the details and arrangements comply with the statutory Regulations and with the requirements of the edition relevant to the year of build of the IEE Regulations for the Electrical and Electronic Equipment of ships or the equivalent standards listed in Merchant Shipping Notice No. M1672. Generators, motors and transforming equipment, intended for services in connection with the propulsion and safety of the ship, are to be inspected during construction. The works and shipboard tests are to be witnessed. Machines and transforming equipment of less than 100kW may be accepted without survey at the manufacturing works if satisfactory maker's test certificates are supplied detailing the results of tests to determine the operating characteristics, temperature rise, insulation resistance and dielectric strength.

NUMBER AND AVAILABILITY OF MAIN GENERATORS

Regulation 54(1) requires at least two main generating sets which together can supply the power for maintaining all normal operational and habitable conditions, that is to say the full design load including non-essential consumers. These generators should be available for use at all times when the ship is at sea, or is about to put to sea. Regulations 54(3, 4, 5 and 6) present the requirements for ships built after 1st September 1984 where one of the generating sets is out of action. Ships built before the 1st of September 1984 shall be provided with at least two generating sets whereby if one set is out of service the remaining set or sets will be capable of supplying those services essential for the propulsion and safety of the vessel after load shedding. Where the services required by Regulation 53(3) are intended to be supplied by a single generator in operation, the main generators should be arranged so that another selected machine is automatically started and connected to the main switchboard if the generator in operation is overloaded or fails. Shaft generators which are not in accordance with the operational requirements of

Regulation 54(3) (*i.e.* cannot maintain services regardless of the speed and direction of rotation of the propulsion machinery or shafting) may be used to supply the main switchboard at sea, provided they are additional to those generators forming the main source of electrical power which are in conformity. The arrangements should be such that the additional generators can easily be disconnected from the main switchboard, without loss of normal supply, as might be necessary when entering and leaving port and in areas where navigation demands special caution.

SAFEGUARDING GENERATORS ETC. AGAINST FLOODING

Where reasonably practicable, installations and equipment should be designed and located such that services which are essential for the propulsion and safety of the ship will not fail in the event of partial flooding of the ship's machinery space. Due consideration should be given to space limitations and the intended service of the ship.

TRANSFORMING EQUIPMENT

Any transforming equipment supplying an electrical system referred to in Regulation 54 should be arranged to ensure the same continuity of supply as that required for generating sets by that regulation. When transformers, rectifiers, etc. are used to supply the services required by Regulation 54, alternative supplies should be available by:-

- an alternative supply from another working transformer/rectifier or generator;
- the provision of a spare transformer/rectifier;
- in three phase circuits, the provision of a spare single phase unit if single phase transformers/rectifiers are used; or
- in three phase circuits, the provision of single phase transformer/ rectifier units connected and of such capacity that the required load can be met with one unit out of service (Open delta connection).

PROTECTION

Short circuit protection must be provided in the primary circuit. Overload protection should also be provided for the primary circuit but the overload device may be located in the secondary circuit of transformers having only one secondary winding.

LOAD SHEDDING

The load shedding, or other arrangements provided, should not interfere with the supply to electrical services which are essential for the propulsion and safety of the ship and which must be maintained after shedding non-essential loads. The essential services include those auxiliaries necessary for propulsion, bilge and fire pumping, navigation lights, essential lighting, steering gear, navigational aids required by the Merchant Shipping Regulations, communication and alarm systems required for safety, watertight doors and engine and boiler room ventilation. The order in which non-essential services are disconnected may be arranged to suit the owner's convenience. Circuits supplying Bow/Stern thrusters should not normally be included in the load shedding arrangements.

MAIN GENERATORS

In general, the location of the main generators in relation to the main switchboard, as referenced by the "same space" referred to in Regulation 54(2) means the same main fire zone and watertight compartment. An environmental enclosure or a subdivision bulkhead or other division which is not required by the regulations, is not to be considered as separating the equipment.

PROTECTION

Where the distance between the generator and associated circuit breaker is larger than usual and in all cases where the generator capacity is large (1500 kW approx), it is recommended that protection is provided which will de-excite the generator in the event of a severe internal fault or of a fault between the generator and its circuit breaker.

MAIN SWITCHBOARD

Circuit breakers or fuses of suitable rating and characteristics could be accepted as a suitable means for subdivision of switchboards required by Regulation 54(6). The advantages of such an arrangement should also be taken into consideration in the design of switchboards where the installed electrical power of the generating sets is less than 3 megawatts.

EMERGENCY AND TRANSITIONAL SOURCE OF ELECTRICAL POWER (REGULATION 56)

LOCATION

The generators, batteries, transformers, switchgear etc, forming the emergency source of power should be sited as per Schedule 7 to Merchant Shipping Notice MSN 1698 (M) so that flames, heat or smoke issuing from skylights, doors or other openings in the event of a fire in the machinery space will not impede the safe operation of the equipment or render the space inaccessible or uninhabitable to necessary personnel.

CAPACITY AND SERVICES SUPPLIED

An emergency generator provided in compliance with Regulation 56 should be of sufficient capacity to supply, simultaneously, those services stipulated in Schedule 7 to Merchant Shipping Notice MSN 1698 (M) as appropriate together with the drencher pump when required by the MS (Fire Protection) Regulations 1998, including starting currents. Paragraph 4 of Schedule 7 to Merchant Shipping Notice MSN 1698 (M) states that emergency generating sets may be used to supply services other than emergency supplies exceptionally for short periods. The primary intent of this relaxation is to allow the use of the emergency source of power when starting the ship's plant from the dead ship condition. Arrangements which would require the emergency generator set to be operated for purposes other than supplying emergency services or testing or the purpose mentioned above, should not normally be accepted.

PERIODIC TESTING

To ensure that the means provided for periodically testing the emergency source of power are fully effective, the arrangements should include facilities for putting the source of power on load by supplying the emergency services. In the case of highly rated turbo charged generating sets, the load should be sufficient to prevent an excessive build up of carbon in the prime mover, which may occur due to operation at light load.

BATTERIES

EMERGENCY BATTERIES

Batteries should be sized according to the maximum design load, in ampere-hours, for which they are required to maintain an emergency supply in compliance with paragraph 13 of Schedule 7 of Merchant Shipping Notice MSN 1698 (M). Surveyors should be satisfied that the capacity is sufficient to ensure performance of the required duty. A Certificate of Inspection by a battery maker of repute, or his agent, regarding the satisfactory condition of the battery, may be accepted. Such Certificates should be supplied to the ship. Discharge of any independent accumulator batteries provided in compliance with paragraph 13(3) of Schedule 7 of Merchant Shipping Notice MSN 1698 (M) should be indicated at the appropriate control station.

TRANSITIONAL SOURCE OF EMERGENCY ELECTRICAL POWER

The charging arrangements provided for the transitional source of emergency electrical power, together with any independent battery provided should be supplied from the emergency switchboard.

BATTERY COMPARTMENTS

Access to battery compartments should, wherever practicable, be arranged from the open deck. Electrical equipment in compartments for the normal "vented type" battery should comply with the IEE Regulations. Where all batteries in the compartment are of the "sealed type", the alternative arrangements described in the IEE Regulations 1990 edition as amended may be accepted, provided all such requirements are met. Some ventilation should always be provided for the space where the batteries are located.

ELECTRICAL PROTECTION OF CABLES

In some circumstances it may be impracticable to provide overload and short circuit protection for cables connected to batteries *e.g.* within battery compartments and in engine starting circuits. Unprotected cable runs should be kept as short as possible and the cables should be "Short Circuit Proof", *e.g.* single core with a sheath or additional insulated sleeve over the insulation of each core.

INVERTERS

Where central inverters (from dc to ac irrespective of size) are used to supply a number of emergency loads, particular attention should be paid to reliability. Equipment of proven design in the marine environment should be selected wherever practicable. Works tests of completed equipment should be witnessed.

EMERGENCY SWITCHBOARD

For the purposes of paragraph 9 of Schedule 7 of Merchant Shipping Notice MSN 1698 (M) an environmental enclosure within the main boundaries of the space does not provide separation between the emergency generator and the emergency switchboard. Circuits should be arranged so that the emergency lighting is of the maintained type, *i.e.* is continuously supplied and forms part of the normal lighting system. Because difficulties can be experienced with the start of certain lamp types (*e.g.* fluorescent tubes) at low temperatures, lamps connected to emergency circuits should normally be of the filament type. Evidence should be provided to confirm that the performance of other types will be satisfactory down to -15°C. Supplementary emergency lighting required for Ro-Ro ships should normally be supplied from the emergency switchboard and the light fittings should be approved in accordance.

PROTECTION OF NON-EMERGENCY SERVICES

The automatic disconnection of non-emergency circuits from the emergency switchboard required by paragraph 11 of Schedule 7 of Merchant Shipping Notice MSN 1698 (M) should be provided where the capacity of the emergency source of power is insufficient to supply the connected emergency and non-emergency services simultaneously. Load shedding or an arrangement interlocked with the interconnector feeder required by Schedule 7 of Merchant Shipping Notice MSN 1698 (M) would be acceptable.

INTERCONNECTOR TO MAIN SWITCHBOARD

The interconnector feeder arrangements required by regulation should be of sufficient capacity to supply all the emergency services and all the non-emergency services which can be simultaneously connected to the emergency switchboard. Clear instructions for manual operation of the emergency source of power and emergency switchboard should be displayed.

STARTING ARRANGEMENTS FOR EMERGENCY GENERATING SETS

Where both the sources of stored energy required by paragraph 12 of Schedule 7 to Merchant Shipping Notice MSN 1698 (M) are accumulator batteries a single common starter motor and associated switch can be accepted. The arrangements should ensure that only one battery at a time can be brought into service. Separate charging arrangements, supplied by the emergency switchboard, should be provided for each battery. A sign should be clearly displayed adjacent to the generating set warning that it is arranged for automatic starting.

SAFEGUARDING OF ESSENTIAL AND EMERGENCY SERVICES

ESSENTIAL AND EMERGENCY SERVICES

Where remote control is provided, local control should be unaffected by a fault, including a cable fault, when local control is selected.

EMERGENCY SERVICES

Where changeover facilities which affect the operation of emergency services are provided (*e.g.* battery isolating switch), the arrangement should be such as to minimise the risk of the devices being left in a position which would prevent the emergency services being supplied when required, *e.g.* locking facilities.

COMPUTERISED EQUIPMENT OR PROGRAMMABLE ELECTRONIC SYSTEMS

If the failure of one particular hardware or software channel could cause the installation to fail in a dangerous way likely to lead to an accident, then at least one additional method of achieving the required safety level following a single failure should be provided. The most usual means of satisfying this requirement is the provision of additional/diversified hardware/software or the addition of nonprogrammable hardware, *e.g.* a "hard wired" circuit.

PUBLIC ADDRESS SYSTEM

A public address system, required by other regulations is also a part of the internal communications system required in an emergency by paragraph

13(3)(a) of Schedule 7 of Merchant Shipping Notice MSN 1698 (M). The loudspeaker arrangements should be such that all public spaces including open decks, public rooms, control stations, alleyways and stairways will be served by more than one circuit.

The circuits should be as widely separated as practicable and cables should preferably be routed outside the zone (*i.e.* above deck), but protection should still be provided. The system should include automatic change-over upon fault to a reserve amplifier. If oscillators or similar electronic means are used to generate emergency warnings and signals at least two oscillators with associated control devices should be provided.

In any system used for emergency signals the oscillator or microphone originating the signal should have facilities to override all other inputs to the system and to override the user volume controls provided. In such systems the circuits to the alarm devices (*e.g.* bells, loudspeakers, etc) should be protected by fuses or other appropriate means so that damage to one circuit will not prevent operation of the remainder of the system. Cabling for emergency alarms and public address systems fitted in Ro-Ro passenger ships on or after 1st July 1998 is to be of a type approved by the Administration.

The MCA considers that cables manufactured in accordance with the standards set out of these Instructions or their equivalent, will be acceptable. Such cables and wiring should, as far as is practicable, be routed clear of galleys, laundries, machinery spaces of Category A and their casings and other high fire risk areas unless serving those spaces.

Where practicable, all such cables should be run in such a manner so as to preclude their being rendered unserviceable by heating of the bulkheads that may be caused by a fire in an adjacent space. All areas of each fire zone should be served by at least two dedicated loops sufficiently separated throughout their length and supplied by independent amplifiers.

CONSTRUCTION OF CABLES

To meet the flame retardant characteristics required by Regulation 58 cables should be at least in accordance with BS 4066: Pt 1: 1995 or IEC 332-1: 1979. To meet the fire resistant characteristics required by Regulation 58 cables should be either of the mineral insulated metal covered type in accordance with BS 6207: 1995 or of a type which meets test symbols CWZ (950°C for 3 hours) as defined in BS 6387: 1994 and is suitable for shipboard use.

Alternatively, compliance with IEC 331 is acceptable as a minimum standard. In a ship constructed on or after 1 September 1984 the insulation of any distribution system that is not earthed should be continuously monitored by a system capable of giving audible and visual indication of abnormally low insulation values.

INSTALLATION OF CABLES AND EQUIPMENT

Main and emergency cable runs and the equipment which they supply should be separated as widely as practicable. Where main and emergency supplies are required for a particular service, *e.g.* the sprinkler pump, the cables should take differing routes, as far as is practicable.

PRECAUTIONS AGAINST FIRE AND FLOOD

Emergency cable runs should be arranged so that a fire or flood in one watertight compartment or fire zone does not affect emergency services in other watertight compartments or fire zones. Cable penetrations through bulkheads which are required to be fire resistant and/or watertight should be made using cable transits which have been approved for this purpose.

PRECAUTIONS AGAINST FIRE AND HIGH TEMPERATURE

Cable runs should, as far as practicable, avoid routes which pass over or near the top of diesel engines and oil-fired equipment, or near to hot surfaces, *e.g.* diesel engine exhaust systems. Where there is no alternative route, cables should be protected from heat and fire damage. Such fire protection may be in the form of a steel plate or trunk, due account being taken of the effects on cable rating, if appropriate. Consideration should be given to the arrangements of bunched cable runs to ensure that their flame retarding characteristics are not impaired.

Note: The use of unsuitable paints, trunking, casings, etc. may significantly affect the fire propagation characteristics of cables. Builders proposals for fire stops in long runs of bunched cables should be considered, unless the cables are totally enclosed in cable trunks. Long cable runs may, as a guidance figure, be taken as those over 6m vertical, 14m horizontal. Where cables are installed in totally enclosed cable trunks, derating may be necessary due to lack of ventilation.

PRECAUTION AGAINST COLLISION

All cables and associated equipment supplied from the emergency source of power and, as far as practicable, main runs of cables supplied from the main source of power, should be kept within the B/5 line. Where supplies to emergency services cannot be kept within this line additional precautions, *e.g.* duplication of supplies, should be taken to protect the services against loss of supply.

CABLE JOINTS

Joints in cables may be accepted for repair or replacement purposes and also, exceptionally, to facilitate installation in new ships. Details of the jointing system, which should be a well established method, should be specially considered. Cable joints should not normally be accepted in hazardous areas, unless they are part of an intrinsically safe circuit.

ELECTRICAL PROTECTION

Protective devices, either individually or in combination, should be capable of handling the maximum short circuit current that may occur at the point of installation, allowing for all the generators that can be running and connected, and including the contribution from all the motors that can be expected to be running.

The shipbuilder's calculations, indicating compliance with the foregoing, should be considered. The operating times of protective devices provided for any circuit should be such that faults will be isolated before the supply cable of the circuit has been permanently damaged. It is recommended that, where practicable, isolating devices which are not designed to break current should be protected against inadvertent or malicious operation when the circuit is carrying current, *e.g.* by interlocking. As a minimum, a notice warning that the device should not be opened under load, should be displayed at the operating position.

The arrangements of circuits and protection supplying emergency services should be such that, as far as practicable, a fire or flood or other casualty in one fire zone or one watertight compartment will not affect the emergency services in other compartments.. An overload alarm should be provided for each circuit where overload protection is omitted.

STEERING GEAR CIRCUITS

Cables and motors for steering gears should be protected against overcurrent of not less than a short circuit. This should be interpreted as meaning that protective devices should not operate at less than 200% of rated motor current. If, however, switching arrangements are such that more than one steering gear motor can be connected to one feeder, its over-current protection is to be similarly based on at least 200% of the sum of the rated currents of all the motors that can be connected to the feeder. The control circuits of electrically controlled steering gear should be connected to the power circuit supplying the steering gear.

ELECTRIC LIFTS

Arrangements should be made for an inspection of lifts, other than those used exclusively for goods, to be carried out by the lift maker or another competent organisation. Satisfactory certificates or reports should be supplied before acceptance. Such lifts should be provided with acceptable means of escape for the occupants for use in the event of the failure of the primary means of escape. One lamp in each lift car, the emergency lighting in the trunk and the alarm system should be supplied from the emergency source of power.

Attention is drawn to Marine Guidance Note MGN 56 (M+F): "Accidents Involving Personnel Lifts".

PORTABLE AND TRANSPORTABLE ELECTRICAL APPARATUS

The effect of Regulation 59(2) is to classify all spaces of the ship, except dry areas of accommodation such as public rooms, cabins and alleyways, as spaces where special provision should be made to ensure that the danger of electric shock is reduced to a minimum. The voltage of electrical supplies to portable and transportable electrical apparatus in all spaces, except dry areas in the accommodation, should be as low as is practicable for the application. Attention is particularly drawn to the recommendation that 24 volts is used for hand lamps.

Where the supply exceeds 55 volts the use of Class I appliances is recommended in preference to Class II. (Class I appliances are provided with facilities for earthing non-current carrying parts.) (Class II appliances have double insulation and/or reinforced insulation throughout and are without provision for earthing.)

It is recognised that the limits of voltage recommended in the relevant Appendices 2 and B of the IEE Regulations 1972 and 1990 respectively are not always practicable for portable and transportable apparatus such as submersible pumps, deck scalers, refrigerated containers etc. In these cases, where 3 phase supplies up to 500 volts may be involved, the additional precautions in 7.31.3.1 and 7.31.3.2 below, or a combination are recommended:-

- circuits which monitor the continuity of the earthed connections and automatically disconnect supply on loss of earth continuity. This arrangement will not, however, be effective when Class II apparatus is used; and
- each socket outlet or group of socket outlets supplied through a high sensitivity residual current circuit breaker (RCCB) (formerly termed high sensitivity current operated earth leakage circuit breaker). For this method to be fully effective the supply must be earthed at one point. In ships with unearthed systems double wound isolating transformers with one point of the secondary winding solidly earthed should be used. Particular attention is drawn to the need to select an RCCB which is resistant to the marine environment, *e.g.* vibration, salt atmosphere etc.

The supply arrangements to refrigerated vehicles or containers on a trailer etc should ensure that all wandering leads are kept clear of the vehicle decks and the need for extension leads is eliminated. Where electric welding apparatus is involved, attention is drawn to Appendix B of the 1990 IEE Regulations.

SUPPLIES TO LIFEBOATS AND SIMILAR CRAFT

The electrical supply connections, if any, from the ship to any lifeboat or similar craft should be at a voltage not exceeding 55 volts DC or 55 volts RMS AC.

SPACE HEATERS

When the installation of space heaters is undertaken, it should be ensured that there will be no risk of heat causing burning or scorching to the surrounding or adjacent material. It is recommended that electric space heaters are provided with thermostats to reduce the risk of overheating.

SPECIAL CATEGORY SPACE VENTILATION

Where regulation requires that any loss or reduction of ventilation capacity is indicated on the navigating bridge, systems based on sensing the current drawn by the fan motor are preferred, as the reliability of the flow switches previously used has been found to be unsatisfactory in marine conditions. Means which indicate that the motor is running, for example an auxiliary contact on the fan motor controller, can be accepted as a minimum requirement if open/close indication of the fire dampers in the respective vent trunking is provided near the motor running lights in the wheelhouse.

ELECTRICAL EQUIPMENT IN HAZARDOUS AREAS AND SPACES

DEFINITIONS

For the purposes of these Instructions, a hazardous area is defined in Regulation 2(2) and includes:-

- those parts of special category spaces where electrical equipment is required by Regulation 60(3) to be certified for use in explosive petrol/air mixtures, unless dangerous goods are to be carried. Decks or platforms with openings of a sufficient size to permit penetration of petrol gases downwards are not considered to be decks on which vapour accumulates;
- when vehicles are carried on multiple levels in special category spaces, above the bulkhead deck the zones regarded as hazardous will be extended in the vicinity of the deck end and openings as shown at A to F of figure 7.35.1.2. Item B in the figure, shows the arrangement when there is a hole in a solid deck, but where the dimension 'd' is more than 1.1h + 2 metres, a mirror image of item A may be used. Uniformly perforated decks from which at least 40 per cent of the area has been removed may, if they are self-draining, be left out of account (item E). Other perforated decks should be

considered as item F. For special category spaces situated below the bulkhead deck any electrical equipment, if fitted, is to be of a type suitable and certified safe for use in explosive petrol and air mixtures;

- cargo spaces as defined in Regulation 60(4);
- ventilation trunks as defined in Regulation 60(5);
- those spaces defined in Regulation 60(6) (Dangerous Goods);
- battery compartments containing ventilated type batteries;
- compartments in which gas burning or welding cylinders are stored;
- compartments in which vapour from heated fuel oil may accumulate in normal circumstances, *e.g.* non-mechanically ventilated spaces containing any part of a heated fuel installation, including tanks; and
- other compartments in which substances which might give off flammable vapours or gases are stored such as paint stores.

TYPE OF EQUIPMENT

Electrical equipment for use in the of these Instructions, with respect to petrol and heated fuel oil, should be certified to at least apparatus group IIA and temperature class T3 as defined in BS 5345: Pt 1: 1989. The following types of equipment may be accepted in these spaces:

- Intrinsically safe Ex i
- Flameproof Ex d
- Increased safety Ex e (except Motors)

Equipment which is required to be enclosed and protected to prevent discharge of sparks by Regulation 60(3) should have a minimum degree of ingress protection IP 55 in accordance with BS EN 60529: 1992. Where this is not practical as with fire detector leads and gas detectors, the equipment should be certified for use in hazardous areas. Equipment which is required to be enclosed and protected to prevent discharge of sparks by Regulation 60(3) should have a minimum degree of ingress protection IP 55 in accordance with BS EN 60529: 1992. Where this is not practical as with fire detector leads and gas detectors, the equipment should be certified for use in hazardous areas. The appropriate apparatus group, temperature class and degree of ingress protection for electrical equipment for use in spaces of these Instructions will depend on the substances to be carried as cargo or stored respectively. To ensure that electrical equipment is suitable for use with all the flammable dusts, gases or vapours to which it may be exposed, electrical equipment, certified to the highest standard, (*i.e.* group IIC, temperature class T6 and degree of ingress protection IP 6X) should be installed. The following types of equipment may be accepted in these spaces:

- Intrinsically safe Ex i
- Flameproof Ex d
- Increased safety Ex e (luminaires only)

Attention is drawn to the special requirements for the carriage of dangerous goods in Regulation 103 of the Merchant Shipping (Fire Protection) Regulations 1998; SI 1998/1012. Electrical equipment in the whole of the space in which the goods are carried, together with the associated ventilation trunks, will normally need to be certified to the highest standards mentioned above, unless the equipment and associated cables are capable of being electrically isolated by the removal of links or the operation of lockable switches.

Electrical equipment for use in the spaces of these Instructions should be certified to at least apparatus group IIC, temperature class T1. Equipment should be confined to luminaires of flameproof type Ex d, or increased safety type Ex e and intrinsically safe circuits. In order to comply with Regulation 83, electrical equipment should not be installed in the spaces of these Instructions.

Only self-contained battery operated lamps or torches of a type that is certified for use in the flammable dusts, gases or vapours to which they may be subjected should be provided for use in hazardous areas and spaces. No facilities for connecting portable electrical equipment should be provided in hazardous areas and spaces.

CABLES

Intrinsically safe circuits should not be run in the same multicore cable as non-intrinsically safe circuits. Different intrinsically safe circuits should not be run in the same cable without special consideration.

The electrical parameters (capacitance, inductance and resistance) of cables for intrinsically safe circuits should comply with the certification documents. Cables for intrinsically safe circuits in new ships should be segregated from all non-intrinsically safe cables by at least 50 mm throughout their length.

In existing ships, arrangements without cable segregation can be accepted provided that either the intrinsically safe cables or the non-intrinsically safe cables are armoured (wire, braid or tape) or metal sheathed. Cable for non-intrinsically safe circuits in the hazardous areas should be either:-

- of the mineral insulated metal covered type;
- protected by electrically continuous metal sheathing or metallic wire armour, braid or tape; or
- enclosed in screwed heavy gauge steel solid drawn or seam welded and galvanised conduit. The conduit should be made gas tight with respect to hazardous areas.

APPROVAL - ELECTRICAL INSTALLATIONS

The following aspects should be considered for inclusion in any approval programme:-

- Cables:
  - Specifications, including flame retardant standard.
  - Location of principal runs of main cables.
  - Location of all emergency cable runs.
  - Details of fire stops in bunched cable runs over 6m vertical; over 14m horizontal in length.
  - Detail arrangements including segregation and marking of High Voltage cables (where applicable).
  - Details of cable transits for watertight and for fire resisting bulkhead penetrations.

LOAD SCHEDULES

Main system, indicating essential services and operational mode (*i.e.* at sea, manoeuvring, harbour, etc). Emergency system, indicating required services and any non-statutory services.

Starting conditions for motors more than 25% of the generator rating are to be stated. Arrangements for the disconnection of non-essential services from the main switchboard and, where applicable, non-emergency services from the emergency switchboard.

Generation and distribution systems (main and emergency):

- Fault level calculations.
- Short circuit capacity of circuit breakers, fuses and bus-bars.
- Type and rating of circuit breakers and fuses.
- Details of protection-overload, reverse power, preference tripping and short circuit.
- Generator paralleling arrangements.
- Generator control. AVR and metering details.
- Earth indication system details.
- Earthing of current transformer secondary windings.
- Interlocking arrangements.
- Switchboard construction (dead front if above 250 volts DC or 55 volts AC).

Emergency stops:

- Means and location of stops for ventilating fans, oil pumps and overboard discharges in way of lifeboat lowering positions.
- Interconnection arrangements between main and emergency switchboards.

Main generators:

- Arrangement drawings showing outline and detail of construction including AVR and exciter, if fitted.
- Position in ship with regard to flooding level.
- Location with respect to switchboard, fire zones and watertight bulkheads.

Emergency generators:

- Arrangement drawings showing outline and details of construction including AVR and exciter.
- Position in ship and with respect to the emergency switchboard and main machinery spaces.
- Auto-start arrangements.

Batteries for essential and emergency services:

- Battery type, manufacturer, service and rating.
- Indication of discharge at main switchboard or in the machinery control room.
- Separation of alkali and acid types.
- Instrumentation and charging arrangements.

Steering gear:

- Details of supply, control circuits and cable runs.

Drencher pump, sprinkler pump:

- Motor power supply, cable runs and protection arrangements.

Main and emergency lighting:

- Position and rating of luminaires
- Provision of at least two separate circuits in each fire zone, main propelling machinery space, saloons, etc.
- Provision of illuminated signs.
- Emergency lighting of maintained type.
- Emergency lighting at boat stations.
- Details of any proposal to use fluorescent luminaires for outside emergency lighting (should normally be of filament type).
- Arrangements of dimming circuits in public rooms etc, if fitted.
- Type, disposition and supply arrangements of supplementary emergency lighting required by Regulation for Ro-Ro ships and which should be in accordance with 10.3.
- Type, disposition and supply arrangements (if applicable) of low location lighting required by Regulation for passenger ships carrying more than 36 passengers and which should be in accordance with the requirements set out in Survey of Fire Protection in Ships - Instructions for the guidance of surveyors.

Navigation lights, access opening indicator lights:
- Supply arrangements including control and alarm panel for navigation lights.
- Arrangements (including details of circuits), limit switches and indicator panel for the access opening indicator lights required by Regulation 86 for Ro-Ro ships.

Lifts:
- Automatic decking arrangements (except those for goods only).
- Emergency escape arrangements (including safety interlocking).
- Supplies to lighting in lift car and trunk (main and emergency) and to alarm.

Shore supply:
- Location and details of connection box(es).

Transformers:
- Protection (electrical) of primary windings.
- Redundancy of transformers for power and lighting.

Plugs and sockets:
- Voltage of sockets for portable and transportable equipment.
- Different socket arrangements for each different voltage.
- Supply arrangements to refrigerated containers etc. where applicable.

Converters:
- Details of converters (rectifiers and inverters) where used for emergency supply purposes.

Automation:
- Details of converters (rectifiers and inverters) where used for emergency supply purposes.

Automation:
- All electrical equipment in hazardous areas to be listed with details of explosion protection and ingress protection, (IP).
- Copies of certification documents for all such equipment.
- Full details of all cables in hazardous areas and the separation of intrinsically safe cables from all other cables.
- For ventilation fans, details of "sparkproof" construction and details of access to fan and motor for inspection and maintenance.

Public address and alarm systems:
- Location of loudspeakers, alarm bells, etc.
- Power supply arrangements for each system.
- Distribution circuit arrangements for each system.
- Redundancy of amplifiers, oscillators, controls etc.

Spares and tools:

Lists in sufficient detail to confirm that Regulation 61 will be complied with.

RE-SURVEY - ELECTRICAL INSTALLATIONS

GENERAL

Surveyors should satisfy themselves that routine testing and servicing of the installation and equipment has been carried out and in particular on essential and emergency services. Casualty investigations have shown that insufficient attention is paid to ensuring that connections and securing devices are tight. Shipowners should check connections at regular intervals so that any wear or other deterioration which could lead to loose parts, overheating or disintegration, is detected and corrected before any serious consequences arise. It is recommended that all current carrying parts and connections should be so constructed and secured as to ensure their continued effectiveness when subject to vibration. Screws and nuts securing current carrying parts should be effectively locked so that they cannot work loose under vibration. Where connections cannot be checked, *e.g.* because of the need to maintain supplies to essential services, shipowners and surveyors are recommended to consider the use of thermographic imaging equipment to detect poor connections by means of generated heat. It should be established that the necessary spare gear is on board and in good condition.

INSULATION RESISTANCE

Surveyors should be satisfied that the insulation resistance value of all cables, switchgear, generators, motors, heaters, other electrical apparatus in power circuits and all lighting circuits (including lighting fittings), are not less than the values given below. Values for essential power and all emergency services should be measured annually and the remainder of the installation within 5 years. Readings presented by the ship's engineers may be accepted at the surveyor's discretion. When it is necessary for values to be measured, they should be taken between all insulated poles and earth. The installation may be sub-divided to any desired extent.

Nominal Voltage of Circuit	Minimum Insulation Resistance In Megohms
Below 50V	0.3
50V to 440V	1.0
Greater than 440V	$\frac{\text{Nominal voltage} + 1}{1000}$

Note: Insulation resistance of intrinsically safe circuits should not normally be measured.

Motors and generators for essential and emergency services The condition of stators, rotors and their windings and the electrical connections, commutators, slip rings etc. should be satisfactory. Main and emergency generators should be seen under working conditions each year.

SWITCHBOARDS, DISTRIBUTION BOARDS AND MOTOR CONTROL GEAR

It should be established that all switchgear is clean and circuit breaker contacts are in good condition. Sealed contacts, which are used in some high voltage circuit breakers, *e.g.* vacuum and sulphur hexafluoride (SF6), need little maintenance and should only be maintained in accordance with the manufacturer's instructions.

The settings of protective devices and fuse ratings should be satisfactory, *e.g.* not excessively high. Particular attention should be paid to ensure that there are no loose connections and parts. Reference is made to 7.37.1.2. In addition, it is recommended that the main and emergency generator circuit breakers should be tested at least every five years. The tests should, as far as is practicable, check the settings of protective devices including over-current, reverse power and preference trips. Simulation by current injection may be accepted.

Where circuit breakers are used to control equipment requiring frequent switching operations, *e.g.* the main generators and bow thrust motors of cross channel ferries, their condition is likely to deteriorate more quickly and a more frequent programme of inspection/maintenance may be required. Consequently, annual inspection and test of these circuit breakers is appropriate and should include the mechanical linkages etc. of units with sealed contacts, *e.g.* vacuum.

ELECTRICAL PROTECTION

It may be appropriate to carry out random checks on the electrical protection devices, *e.g.* fuse rating, MCB setting, overload setting in motor starters etc. Electric welding plant should be in accordance with Appendix B of the 1990 IEE Regulations.

TRANSFORMING EQUIPMENT

Where transformers, rectifiers etc are used to supply essential or emergency services the surveyor should satisfy himself regarding their general condition with particular reference to cleanliness and the tightness of connections and parts.

Where liquid immersed equipment is installed, an up to date certificate from a competent test authority testifying that the electrical characteristics of the liquid are satisfactory for the condition of service should be available to the surveyor at least every 5 years.

ELECTRICAL EQUIPMENT INCORPORATING LIQUID COOLING

Where water circulated air coolers are fitted they should be examined and if necessary subjected to a suitable hydraulic test. Where liquid is in contact with live conductors an up to date certificate from a competent test authority testifying that the electrical characteristics of the liquid are satisfactory for the conditions of service should be available to the surveyor at least every 5 years.

Where suitable test equipment for determining the electrical characteristics of the cooling medium is provided on board, these results may be accepted at the discretion of the surveyor. The alarm and indicators provided, *e.g.* temperature, flow etc, should be tested as far as is practicable.

EMERGENCY SERVICES

It should be established that the operation of equipment for automatically starting and connecting the emergency generator to the emergency switchboard is satisfactory. The emergency services required by the Regulations *e.g.* lighting, pumps, miscellaneous services etc, should be seen under operating conditions each year.

EMERGENCY BATTERIES

It should be established each year that the battery and associated compartment or container is clean and in good condition and that ventilation arrangements are in good order. Any battery cell suspected of being defective should be tested for capacity and if it fails to give 80% of its rated capacity it should be considered unfit for further service.

A Certificate of Inspection by a battery maker of repute regarding the satisfactory condition of the battery may be accepted by the surveyor. Such certificates should be supplied for inclusion in the records of the ship. The arrangements for automatic change-over from mains supply to battery supply required by the Regulations should be checked each year.

CABLES

It may be appropriate, depending on the age and condition of the installation to examine a sample of cables to ascertain the general condition of the whole installation. Opening up of trunks etc should not normally be necessary unless examination or unsatisfactory insulation resistance readings indicate that it is required for further investigation. Particular attention should be paid to cables in areas of high ambient temperatures, *e.g.* near exhaust systems of diesel engines.

ELECTRIC LIFTS

- Surveyors should be satisfied that electric lifts are maintained in good condition.

- Inspection and maintenance of lifts should be carried out by the lift makers or other competent organisations every six months and certificates or reports made available to the surveyor. If such a certificate or report is produced or some equally satisfactory arrangement is made, surveyors need not insist on the opening up of working parts or dismantling of safety devices etc, for inspection. Attention is drawn to Marine Guidance Note MGN 56 (M+F) "Accidents involving Personnel Lifts".

HAZARDOUS AREAS AND SPACES

The condition of cabling and equipment within hazardous areas and spaces should be examined each year to ensure that it is in a satisfactory condition and that no unauthorised alterations have been made. Particular attention should be paid to the insulation resistance values and electrical protection devices. Intrinsically safe circuits should not normally be subjected to an insulation resistance test.

HIGH VOLTAGE EQUIPMENT

Where electrical equipment operating at above 1000 volts is installed any arrangements which are provided to guard against unauthorised entry of personnel should be checked each year to ensure that they are satisfactory.

9

Ocean, Coastal Processes and Tides

GEOLOGICAL OCEANOGRAPHY

Put simply, geological oceanography is the study of the ocean floor. This branch of earth science is focused on the geological structure and topography of the ocean floor and the natural processes that create and modify these features. Scientists specializing in this field apply oceanographic principles and techniques to uncover important geological events that have occurred millions of years ago, as well as to predict those likely to occur in the future. In terms of structure, geological oceanographers study the effects of the movement of the earth's lithospheric plates, which results in sedimentary deposits and the formation or destruction of the ocean floor. One notable area of concentration impacted by these processes is the continental shelf. This is the area of land under the ocean just off the shore, and extends to the point where the shelf slopes and descends to the abyssal plain of the ocean floor, the depth of which varies across the globe.

By measuring the depth of the shelf at specific locations, scientists can predict the pattern of movement of water along the coastal shore, as well as the height of waves that can be expected to initiate offshore and eventually reach land. Studying variations in the continental shelf also provide clues to help estimate fluctuations in sea levels in various geographic regions, as well as the impact of tropical storms. There are numerous secondary disciplines that fall under the science of geological oceanography. For instance, paleoclimatology is the study of deep-sea beds to construct a timeline of climatic change as it relates to shaping and reshaping the Earth, as well as the health and biodiversity of past and present living organisms.

Plate tectonics is the study of the earth's tectonic plates, which reside on the planet's lithosphere and are triggered to migrate due to heat being generated from the underlying upper mantle of the earth's crust, or asthenosphere. Many sub-disciplines of geological oceanography yield even more specialized areas of study. For example, seismology, which is devoted to the study of earthquake activity, is a subcategory of plate tectonics.

Geological oceanography scientists use a wide variety of tools and equipment in their work. For example, in addition to satellite imagery, researchers utilize highly advanced instruments to map the ocean floor, such as high-resolution swath bathymetry survey systems and side-scan sonar and acoustic profiling equipment. Many scientists also participate in deep-sea exploration programs, such as the Ocean Drilling Program, an international project funded and operated by the US National Science Foundation and 22 global partners that make up the Joint Oceanographic Institutions for Deep Earth Sampling.

TOOLS AND TECHNIQUES

Many different tools are used by geological oceanographers. For example, the structure and topography of the ocean floor are studied through the use of satellite mapping, which measures the level of the ocean surface to estimate the shape of the ocean floor. A detailed study in 1978 by the National Oceanic and Atmospheric Administration involved the Seasat satellite which produced frequent, short pulses of microwave radiation to measure the level of the surface of the sea with great accuracy. Underwater mountains and valleys cause subtle variations in Earth's gravitational field.

The stronger gravity near high massive formations attracts more water molecules, raising the level of the ocean slightly. Similarly valleys on the ocean floor produce weaker areas of gravity, so the level of the ocean will be lower. Using this technique, a complete survey of the ocean floor was accomplished. Seismic techniques are used to measure the subsurface structure. This type of study is carried out by teams of two ships: one fires an explosive in the water and the other uses sensitive instruments to record the sound waves as they reach the second ship.

Some waves travel directly to the second ship; others travel to the ocean floor, are refracted within the layers of sediment, and then travel to the second ship. By measuring the time it takes for the energy to arrive and the distance between the boats, the thickness of sediments and other features can be determined. Structures may also be analyzed by studying natural earthquake waves that travel through deeper oceanic rocks and may be recorded at stations around the world. Geological oceanography and coastal geology are closely related yet distinct disciplines.

Geological oceanographers study the rocks and structure of the ocean floor, the oceanfloor sediment that covers them, and the processes that formed them. Coastal geologists focus on these structures and processes in a coastal environment. Other ways oceanic sediments are studied are by dredging processes and deep-sea exploration projects such as the Ocean Drilling Program which obtains samples of seafloor sediment from the entire world. The ocean sediments are found to consist of rock particles and organic remains whose compositions depend on depth, distance from continents, and local variations such as submarine volcanoes or high biological activity. For example, clay minerals, which are formed by the weathering of continental rocks, are carried out to sea by rivers and are usually abundant in the deep sea. Thick deposits of weathered rock material, commonly coarser than deep-sea clays, are often found near the mouths of rivers and on the continental shelves.

BATHYMETRY

Bathymetry is the study of underwater depth of lake or ocean floors. In other words, bathymetry is the underwater equivalent to hypsometry. The name comes from Greek âáèõò, deep, and ìåôñïí, measure. Bathymetric charts are typically produced to support safety of surface or sub-surface navigation, and usually show seafloor relief or terrain as contour lines and selected depths, and typically also provide surface navigational information. Bathymetric maps may also use a Digital Terrain Model and artificial illumination techniques to illustrate the depths being portrayed. Paleobathymetry is the study of past underwater depths.

Measurement

Originally, bathymetry referred to the measurement of ocean depth through depth sounding. Early techniques used pre-measured heavy rope or cable lowered over a ship's side. The greatest limitation of this technique is that it measures the depth only a single point at a time, and so is inefficient. It is also subject to movements of the ship and currents moving the line out of true and therefore is inaccurate. The data used to make bathymetric maps today typically comes from an echosounder mounted beneath or over the side of a boat, "pinging" a beam of sound downward at the seafloor or from remote sensing LIDAR or LADAR systems. The amount of time it takes for the sound or light to travel through the water, bounce off the seafloor, and return to the sounder tells the equipment what the distance to the seafloor is.

LIDAR/LADAR surveys are usually conducted by airborne systems. Since the early 1930s, and more commonly from the 1940s onwards, the occasional pings of a single-beam sounder might be averaged to make a map. Today, a multibeam echosounder may be used, featuring hundreds of very narrow adjacent beams arranged in a fan-like swath of perhaps 90 to 170 degrees across. The tightly packed array of narrow individual beams provides very high angular resolution and accuracy.

In general the wide swath, which is depth dependent, allows a boat to map more seafloor in less time than a single-beam echosounder by making fewer passes. The beams update many times per second, allowing faster boat speed while maintaining 100% coverage of the seafloor. Attitude sensors allow for the correction of the boat's roll, pitch and yaw on the ocean surface, and a gyrocompass provides accurate heading information to correct for vessel yaw. The Global Positioning System is used to position the soundings with respect to the surface of the earth. Sound speed profiles of the water column correct for refraction or "ray-bending" of the sound waves owing to non-uniform water column characteristics such as temperature, conductivity, and pressure. A computer system processes all the data, correcting for all of the above factors as well as for the angle of each individual beam.

The resulting sounding measurements are then processed either manually, semi-automatically or automatically to produce a map of the area. A number of different outputs are currently generated, including a sub-set of the original measurements that satisfy some conditions or integrated Digital Terrain Models. Historically, selection of measurements was more common in hydrographic applications while DTM construction was used for engineering surveys, geology, flow modeling, etc. Since ca. 2003-2005, DTMs have become more accepted in hydrographic practice. Satellites are also used to measure bathymetry.

Satellite radar maps deep-sea topography by detecting the subtle variations in sea level caused by the gravitational pull of undersea mountains, ridges, and other masses. On average, sea level is higher over mountains and ridges than abyssal plains and trenches. Most surveys of navigable waterways in the United States are performed or commissioned by the United States Army Corps of Engineers, for inland waterways, and the National Oceanic and Atmospheric Administration for oceans. Coastal bathymetry data is available from NOAA's National Geophysical Data Center (NGDC). Bathymetric data is usually referenced to tidal vertical datums.

For deep water bathymetry, this is typically Mean Sea Level but most data used for nautical charting is referenced to Mean Lower Low Water in American surveys, and Lowest Astronomical Tide in other countries. Many other datums are used in practice, depending on the locality and tidal regime. Some occupations or careers related to bathymetry are the study of oceans and rocks and minerals on the ocean floor, and the study of underwater earthquakes or volcanoes.

The measurement and analysis of bathymetric measurements is one of the core areas of modern hydrography, and a fundamental component in ensuring the safe transport of goods worldwide.

Ocean-Floor Bathymetry

The term bathymetry is defined as the depth of water relative to sea level. Thus bathymetric measurements can determine the topography of the ocean floor, and have shown that the sea floor is varied, complex, and ever-changing, containing plains, canyons, active and extinct volcanoes, mountain ranges, and hot springs.

Some features, such as mid-ocean ridges and subduction zones, also called deep-sea trenches are unique to the ocean floor. Bathymetric mapping involves the production of ocean and sea maps based upon bathymetric data.

Bathymetric maps represent the ocean depth as a function of geographical coordinates in the same way topographic maps represent the altitude of Earth's surface at different geographic points. The most popular type of bathymetric maps are ones on which lines of equal depths are represented.

Bathymetric Techniques

For hundreds of years, the only way to measure ocean depth was the sounding line, a weighted rope or wire that was lowered overboard until it touched the ocean floor. Not only was this method time-consuming, it was inaccurate; ship drift or water currents could drag the line off at an angle, which would exaggerate the depth reading. It was also difficult to tell when the sounding line had actually touched bottom. England's Sir John Murray compiled this bathymetric chart of the North Atlantic in 1911. Murray's chart went far beyond American naval officer Matthew Maury's first attempt at bathymetric mapping in 1855.

In addition, Murray's map gave birth to the idea of the Telegraphic Plateau, a submarine land formation from Canada to the British Isles, across which the first transatlantic cable was laid. In the twentieth century, sounding lines were entirely replaced by sonar systems. Sonar, invented during World War II measures distances by emitting a short pulse of high-frequency sound and measuring the time until an echo is heard.

After the war, ships with sonar units attached to their hulls crisscrossed the world's oceans systematically, measuring depth. The data collected made possible complete bathymetric maps of the world's oceans. For the first time, scientists knew what 70 percent of Earth's surface really looked like. Many sonar techniques have been developed for bathymetry. When high-resolution images are desired, an underwater unit may be towed behind a ship, scanning to the left and right with multiple sonar beams.

Furthermore, orbiting visible-light cameras image the bottoms of some shallow waters, while satellite radar maps deep-sea topography by detecting the subtle variations in sea level caused by the gravitational pull of undersea mountains, ridges, and other masses.

The Ocean Floor in Cross-Section

The oceans begins, of course, at the shore, the irregular boundary where the surface of a continent descends first to sea level and then beneath it. If the depth of the ocean is measured along a line drawn straight out from a continental shore, the following sequence of bottom features are typically seen.

- Continental Shelf: For many miles out, the ocean is only a few hundred feet deep and gets deeper quite slowly. This flat, wide margin is found around every continent and is known as the continental shelf. The average width of a continental shelf is 70 kilometers.

- Continental Slope and Rise: The continental shelf ends at a sudden dropoff called the shelf break. Beyond the shelf break, the slope of the ocean floor becomes much steeper, typically a 4° slope, or 70 meters per kilometer. This steep embankment is called the continental slope and is grooved by submarine canyons and gullies. The continental slope is about 16 kilometers wide, on average, and descends to a depth of about 2.4 kilometers. There it ends as the slope moderates to a mere degree or two from horizontal. This gradual zone, which may be several hundred miles wide, is called the continental rise. It is composed of fine-grained continental sediments washed down the many submarine canyons that notch the continental slope.
- Abyssal Plain: The abyssal plain, which is the deepest, most level part of the ocean, is found where the continental rise ends, at a depth of about 4 kilometers. The abyssal plain is dotted with thousands of small, extinct volcanoes called abyssal hills. The abyssal plains of the Atlantic appear smooth because its abyssal hills are buried under a thick blanket of continental sediment, but in the Pacific Ocean basin, which is ringed by trenches that trap sediments before they can spread over the ocean floor, tens of thousands of unburied abyssal hills have been observed. Abyssal hills more than 1 kilometer high are called seamounts, and seamounts with flat tops are called guyots or tablemounts. Guyots are drowned volcanic islands that become submerged due to subsidence of the oceanic lithosphere.
- Mid-Ocean Ridge: Beyond the abyssal plain, which may be several hundred kilometers wide, the ocean floor begins to ascend again with a gentle slope. This area is the flank of the mid-ocean ridge, a long, undersea mountain chain that usually extends down the middle of the ocean. The Mid-Atlantic Ridge, for example, snakes down the middle of the Atlantic most of the way from the North Pole to Antarctica. As the ocean floor climbs slowly toward the center of the mid-ocean ridge its sediment blanket gets thinner and its surface more irregular. Here, the ocean floor is marked by thousand-mile cracks called fracture zones that lie at right angles across the mid-ocean ridge.
- Rift Valley: Along the center of the mid-ocean ridge is the rift valley, a deep V-shaped notch. From this valley, new oceanic crust is constantly being extruded from Earth's mantle by processes not yet fully understood. Twin sheets of fresh, mile-thick crust emerge from the mantle along the sides of the rift valley and flow slowly away from it in opposite directions. In the case of the Mid-Atlantic rift valley, one sheet flows east and the other west, each moving at about

half an inch per year. The older, more distant parts of these growing sheets of crust are gradually covered by sediments, eventually becoming the abyssal plains and continental rises described above. These growing sheets of crust have the effect of forcing the distant continents farther apart, a process called sea-floor spreading.

- Subduction Zones: Beyond the rift valley, the depth sequence described above is observed in reverse: a gently declining slope of ridges and fractures, an abyssal plain, a continental rise, a steep continental slope, a somewhat flat continental shelf, and finally dry land again. This would be the case in crossing the Atlantic Ocean. Elsewhere, however, such neat symmetry is not always found. Some continental margins are characterized by subduction zones, whereby oceanic crust is destroyed beneath the edge of a continental plate. The hypsographic curve shows the amount of Earth's surface at various elevations and depths. Horizontal dashed lines indicate average height of the continents at 840 meters above sea level, and average depth of the oceans at 3,800 meters below sea level. The vertical dashed line marks the division between land and sea at present-day sea level. The disappearance of oceanic crust into subduction zones is occurring primarily along the western, northern, and eastern edges of the Pacific Ocean. These subduction zones are the deepest places in the ocean, varying from 7 to 11 kilometers in depth. Along the length of a typical subduction zone or deep-sea trench, a sheet of old oceanic crust is forced beneath continental crust. The oceanic crust that is forced to submerge or subduct in this manner slides down into the mantle at an angle.

The Hypsographic Curve

The elevation features of the ocean floor—shelves, slopes, rises, plains, ridges, and trenches—are quantitatively summarized by oceanographers, along with the distribution of dry land at different altitudes, in a graph called the hypsographic curve. The hypsographic curve shows what percentage of Earth's surface rises above present-day sea level to a given height, or sinks below it to a given depth.

The curve shows that a small percentage of Earth's surface consists of high-altitude continental mountains, 30 percent of the surface consists of continental lowlands. Continental shelves and slopes account for about another 15 percent, abyssal plains and mid-ocean ridge systems for almost 50 percent, and deep-sea trenches account for a small fraction.

The fact that Earth's surface is comprised mostly of flat plates of continental and oceanic crust, with mountains and deep-sea trenches occurring only along their edges where the plates collide, gives the hypsographic curve its characteristic shape.

SEAFLOOR TOPOGRAPHY

Seafloor topography influences ocean circulation in two basic ways. First, it steers ocean flows. Second, it provides barriers that prevent deep waters from mixing, except within deep passageways that connect ocean basins or in hydraulically controlled overflow regions. This paper explores the impact of both of these processes on ocean circulation. The examples highlighted here were among the broad range of topics explored at a workshop on "Ocean Circulation, Bathymetry, and Climate," held at Scripps Institution of Oceanography in October 2002.

Topographic Steering

Ocean currents cannot pass through ridges or seamounts. At ocean depths that are intersected by topography, currents steer around major topographic features. In addition, particularly at high latitudes, where the ocean is weakly stratified, geophysical flows tend to be vertically coherent due to the Earth's rotation. As a result currents near the ocean surface align in roughly the same direction as deep ocean currents, and consequently often follow contours of constant depth, detouring around the bumps and troughs in the seafloor. Most major currents respond to sea floor topography. The Antarctic Circumpolar Current, the Gulf Stream, and the Kuroshio Extension all steer around ridges and seamounts.

The fronts flow to the south of the Campbell Plateau near New Zealand, and through the Eltanin and Udintsev Fracture Zones in the central Pacific Ocean. Just downstream of Drake Passage, around 60°W, they veer northward around the ridges of the Scotia Arc. To the extent that oceanographic flows are strictly barotropic, they should be steered along contours of constant f/H, where f is the Coriolis parameter and H represents ocean depth. Barotropic theory is often supported by observations. For example, floats in the Atlantic and Pacific Oceans preferentially spread along f/H contours rather than across them, indicating that flow responds to topography. In reality, because the ocean is stratified, and velocities tend to be faster near the ocean surface than at middepth, flow does not literally follow contours of f/H. Gille used float data to examine whether Southern Ocean velocities could be assumed to be equivalent barotropic, meaning that velocities attenuate with depth, with a fixed e-folding scale, Ho.

Thus velocity $v(z) = v(0) \exp(-z/Ho)$. Under this assumption, flow is predicted to follow contours of f/Fo, where $Fo = Ho(1 - \exp(-H/Ho))$. In the limit where the e-folding scale, Ho, is infinite, this is equivalent to assuming that flow follows f/H contours. Since only large-scale topographic features of the sea floor are expected to steer large-scale circulation, topography was smoothed to eliminate variations with length scales less than 100 to 200 km. Gille found that in the Southern Ocean the equivalent barotropic model explains the largest fraction of variance in the flow data when an e-folding depth of about 700 m is assumed. This is consistent with other analyses of the Southern Ocean that have suggested that velocities decrease with e-folding scales between 500 and 1000 m, depending on position within the ACC and computation method.

Research using the Naval Research Laboratory Layered Ocean Model has shown the influence of abyssal layer flow on the upper ocean in numerical simulations. Hurlburt and Metzger demonstrated how topographically steered mean abyssal currents can steer upper ocean currents as the Kuroshio Extension bifurcates in the vicinity of the Shatsky Rise. Surface currents can bend in the direction of the abyssal currents. This is particularly noticeable for northward and southward meanders of zonal currents. Hurlburt and Thompson used the continuity equation in a two-layer model to show in a very clear and direct fashion how abyssal currents can steer upper ocean currents, especially where they intersect at large angles. In a multilayer case, the argument formally breaks down, but the steering effect remains in situations where the barotropic and first baroclinic modes are dominant, such as in western boundary currents or in the ACC, and the flow can be approximated by two layers.

Hogan and Hurlburt described an example of upper ocean-topographic coupling using 1/8° – 1/64° Japan/East Sea versions of NLOM. They noted that 1/32° horizontal resolution is required to produce baroclinically unstable surface currents. At that resolution, baroclinic instability is very efficient at transferring energy from the upper layers to the abyssal layer where the currents are constrained to follow the contours of the bottom topography. The result is a profound change in abyssal circulation compared to the coarser resolution simulations. In particular, the eddydriven deep mean flows are much stronger and occur over most of the basin. A southward abyssal flow along the coast of Korea and an anticyclonic abyssal circulation around a ridge near 39°N, 130°E contribute to the separation of the East Korean Warm Current near 37° – 38°N as a result of the upper ocean-topographic coupling described above. These results demonstrate that the bottom topography in this region is critical for the EKWC to separate from the coast at these latitudes. An experiment that removed the ridge near 39°N, 130°E eliminated the offshore abyssal steering and consequently changed the separation latitude of the EKWC.

The mean pathways of major current systems can also be significantly affected by accurate topographic information. Metzger and Hurlburt studied this in the vicinity of Luzon Strait, which connects the Pacific Ocean and the South China Sea. As the North Equatorial Current bifurcates along the east coast of the Philippines, the northward branch forms the beginning of the Kuroshio. Upon entering Luzon Strait, the Kuroshio intrudes westward into the SCS before retroflecting and continuing its poleward journey along the east coast of Taiwan. Using a 1/16° Pacific Ocean version of NLOM, the authors found that the westward extent of Kuroshio intrusion is highly dependent upon the accuracy of the coastline geometry of the island chain within Luzon Strait. Two small-scale shoals were found to have a significant blocking effect on the Kuroshio. Inclusion of the shoals had two effects on NLOM. First, they narrowed Luzon Strait and thus reduced the westward bending. Second, more importantly, they deflected the inflow angle making it more northward, thus reducing the westward intrusion.

The NLOM simulation with the more accurate coastline geometry agrees more closely with the observational evidence of Qu and Liang et al. Given the importance of very small-scale features in determining the pathways of major ocean currents, accurate representation of sea floor topography is crucial when doing high horizontal resolution ocean modeling. Unfortunately, most of today's topographic databases do not adequately resolve small-scale features. In the case of the island chain within Luzon Strait, significant hand editing using navigational charts from the Defense Mapping Agency was required to obtain realistic coastlines. Thus, higher resolution global bathymetry is needed.

Flow Through Ridges and Gap

Topography matters not only because it steers ocean flows, but also because it inhibits or enhances the mixing and transport of waters from different regions. Constrictions, such as narrow fracture zones or shallow sills, can play an important role in determining how water passes between different parts of the global ocean. For a review of the subject and theory see Whitehead, which also contains a list of important passages and straits around the globe. Examples given here, from several ocean basins, illustrate the importance of such bathymetric features.

Nordic Seas and Atlantic Ocean Interface

The overflow waters that flow from the Nordic Seas through the Denmark Strait and down the slope into the Atlantic basin contribute to the formation of North Atlantic Deep Water. Colder, fresher water forms over the shelf and spills over the Denmark Strait, entraining warmer and more saline Irminger Sea water along the way. Historically, ocean models have had problems with dense water overflows that are crucial for accurate simulation of deep water properties and the overall global circulation budget. Part of the reason for this misrepresentation is that in these older models, the flow is determined mainly by the mixing and not by hydraulic or topographical processes. Recently, some realistic process-oriented numerical studies of the Denmark Strait examined the nature of the flow in this region. Both papers included examples that used bottom-following coordinate systems and realistic bathymetry at a high resolution. The Denmark Strait is considered a wide sill, many Rossby radii in width, and it is v-shaped. Shi et al. found that when using realistic topography, the downstream characteristics of the flow were little changed with the inclusion of detailed bathymetry. They found that a change in the mixing scheme had a more profound effect on the downstream flow and its characteristics. Käse and Oschlies, on the other hand, concentrated on examining the flow before it enters the strait and the effect of the topography on flow. Their conclusion was that the transport of the dense water through the strait is controlled by topography and modulated by the downstream eddy field. The volume transport is limited by the topography, as Killworth and McDonald theorized. They further found that while the heat transport fluctuations are largely due to the change in temperature upstream of the sill, the volume transport remains relatively constant. Thus, it is important to determine correctly sill depths and their extent in order to reproduce such flows properly in models. As an example from a primitive equation model, simulated February temperature and salinity sections at 31°W within the Denmark Straits overflow region. These sections are from a 1/12° HYbrid Coordinate Ocean Model, which uses a hybrid of isopycnal coordinates, terrain-following coordinates, and fixed vertical coordinate systems The hybrid vertical coordinate approach effectively handles such overflow situations. The depth of the sill is one of the controlling factors governing the details of the overflow, and it is necessary to know the bathymetry as accurately as possible.

Large-scale flows can also be extremely sensitive to the sill representation even in coarse resolution climate models. Roberts and Wood showed that in a coarse resolution ocean model, the heat transport northward past Iceland differs by a factor of two depending whether their model includes deep passageways through the sill that separates the Greenland-Iceland-Norwegian Sea from the North Atlantic. The coarse resolution used by Roberts and Wood does not fully resolve the sill features or the narrow hydraulic flow over the sill, but the strong sensitivity to topography indicates that climate change signals estimated from coarse resolution models may depend critically on the representation of seafloor topography in the models.

Mid-Atlantic Ridge

In the tropical Atlantic Ocean near 22°S, 3000 m deep waters to the west of the Mid-Atlantic Ridge are high in oxygen, indicating that they have had relatively recent contact with the atmosphere and as a result may carry signatures of recent climate fluctuations. In contrast, waters to the east of the Ridge are low in oxygen. The Mid-Atlantic Ridge, with a typical depth of 2500 m, separates these waters along most of its length. This is a clear indicator that a fracture zone exists in the ridge.

This is a clear indicator that a fracture zone exists in the ridge. No deep fracture zone can be identified in existing Smith and Sandwell seafloor topography, but a detailed multi-beam survey identified a 3900-m deep sill that could explain the observed oxygen values. Because flow through the ridge is constrained to a narrow region and is hydraulically controlled, mixing within this gap may be strongly elevated compared with mixing elsewhere in the deep ocean. Deep sills, such as this one, are important for our understanding of the ocean's role in climate, because they help determine how deep waters can mix and ultimately how heat is transported through the ocean.

Indonesian Seas

The importance of including the correct bathymetry in an ocean model is also seen in the data presented by Gordon et al.. They discussed the bathymetric barriers within the Indonesian seas that influence the through flow of waters from the Pacific to the Indian Ocean, an important climatic choke point in the circulation of our global ocean waters. While most of the transfer of water is in the upper portion of the thermocline layer, Gordon et al. determined that there is a significant contribution from the deep layers. Some passages are as narrow as 30 km wide and 3250 m deep, such as Ombai Strait.

This strait, they estimated, allows for the passage of 0.6 to 0.8°-106 m3 s-1 of deep water or 10 to 20% of the total 4 to 6°-106 m3 s-1. The bathymetry used by the current generation of ocean models requires much hand editing and relies on ship surveyed depth measurements to define the important passages. While the in situ measurements made by this group have been extensive, they commented that most sills in the Indonesian seas have not been surveyed. The paper also suggests that the thickness of the overflow layer is determined by the sill topography itself and that these details influence the amount of water transported over or blocked in the region of the sill. With the resolution of models increasing, these types of details will need to be included in the bathymetry that the models incorporate into their domains.

Discussion: Requirements for the Future

In the future, ocean modelers are likely to become more acutely aware of the sensitivity of their models to topographic details. The highest resolution global ocean models that are currently in use have grid points roughly every 10 km. These models are barely able to respond to the smallest topographic features currently resolved, which have length scales of 20 to 30 km. Projections for the future suggest that as computer power increases, model grid spacings will decrease to 5 km or less by 2010, and will respond to topographic features that are 10 km or less in length and are likely to require topographic position accuracies of ±2.5 km.

As horizontal resolutions are refined, vertical resolution also increases. Currently the largest deep ocean vertical separations are about 250 m. By 2010, average deep ocean vertical separations will decrease to 75 m or less, and models will therefore be sensitive to errors in bottom depth of 40 m or less. These upcoming requirements for bathymetry with higher horizontal resolution and small vertical errors will pose a challenge to those who gather and archive bathymetric data. Coupled ocean/atmosphere models and models representing long-term climate change are likely to run at coarser resolution, but they will nonetheless require accurate bathymetry as well as good parameterizations of bathymetrically-controlled processes, such as hydraulic overflows, that occur at length scales smaller than model resolution.

PLATE TECTONICS

One of the first steps in performing a scientific investigation is to understand what is already known. The investigation could be posed in a number of ways, but I have settled on asking students to find evidence for the theory of plate tectonics, as explained in their textbook. This approach has worked well in a class of general education majors. With more class time it might be possible to lead students to the discovery that the earth's surface must actually be moving around. The seafloor age data requires it, as do the existence of earthquakes in well defined patterns. Nevertheless, the search for this evidence provides rich investigative opportunities. Furthermore, the real earth is uncooperative enough to provide many complications for deeper investigation.

The first step is to familiarize students with the theory as described in the simplistic cartoons that depict spreading centers, transform faults and subduction zones. A clear understanding of the basic theory is essential if students are to successfully find data in support of it. The "Plate Tectonics Lecture" does an excellent job of animating these cartoons to aid in visualization. In some instances, clay models might also prove useful. The following discussion reviews some of the most important ideas of Plate Tectonics.

Later, examples of interesting studies are shown and discussed. The theory of plate tectonics states that the earth's surface has a number of rigid, brittle segments that move around. The outlines of these plates are easy to see on a world map of earthquakes. Earthquakes occur where brittle pieces of the earth are slipping past one another. When you break a dish, a "crack" occurs and there is a separation of two or more pieces. The noise is analogous to the earthquake and the crack is where relative motion between two pieces takes place. The plate is called of "Lithosphere". Within the lithosphere is the crust and upper portion of the mantle. There are three kinds of boundaries between plates. These will each be discussed below.

Divergent boundaries

These are called "spreading centers" and often occur at "mid-ocean ridges". Two plates are separating, or diverging.

Location A is the location of the spreading center in the middle of the Atlantic Ocean. Notice that the spreading center is higher in elevation. Even before plate tectonics, this long linear feature was described as a "midocean ridge". This is because the lithosphere is hottest at a spreading center, so thermal expansion makes its elevation higher. Hot mantle rocks rise into the space left by the separating plates and form the new oceanic crust. The Mid-Atlantic ridge has many large volcanoes on its flanks. The region is characterized by few earthquakes, most of which are on short "transform" segments.

TRANSFORM BOUNDARIES

Transform boundaries are where the plates slide past one another. The San Andreas Fault is a transform boundary between the Pacific and North American Plates. Transform boundaries are characterized by many earthquakes at relatively shallow depths.

The location between the two spreading centers which is the transform fault. This is the only region where there is relative sliding motion. Each of the offset spreading centers is at a higher elevation, so when the offset spreading centers are joined by a transform fault, there are elevation differences at the boundary.. Boundary segments marked by A and C do not have relative motion across the "fault". Yet the original elevation differences caused by the offset are still present.

These segments are called "Fracture Zones", and can be observed in the Map elevation data in the eastern Pacific Ocean, where they may persist for thousands of kilometers. In map view, the representation of offset spreading centers, joined by a transform fault is shown to the left. The transform fault is indicated by the B, and A and C are the fracture zones, which carry the scar of the old transform fault along as the plate spreads away from the spreading center.

There are several rules that these structures obey:

- The transform fault is at a right angle to the spreading center
- Both spreading centers are spreading at the same rate, unless there is relative rotation between the two plates.
- Over time, spreading centers can move. This happens if one side spreads faster than the other and since the spreading center must be in the center of the two spreading plates, it moves in the direction of the faster spreading plate.
- Over time, spreading centers can become longer or shorter. When this happens, they are often called "propagating rifts". The length of the spreading center increases in the direction of the propagation. This means that the transform fault that connects a propagating spreading center to another spreading center must move too.

CONVERGENT BOUNDARIES

There are a number of important variations to converging boundaries. The collision between India and Eurasia is an example of a plate convergent boundary where two continents collide. The continents are both too buoyant to sink, so their collision causes a thickening of the lithosphere, which is manifested in the Himalayan Mountains and Tibetan Plateau.

When an oceanic plate collides and subducts beneath another oceanic plate, an island arc forms. The islands are caused by the volcanoes created by the subduction process. Behind the island arc, a basin is formed. Often, a small spreading center occurs, caused by the pull of the subducting slab. This is called "back-arc spreading".

COMPLICATIONS

There are many "real world" examples of the simple structures described above. You will be able to find these using the map software. However, there are also many complications, which are beyond the scope of this treatment. One interesting complication occurs at the northern end of the San Andreas Fault. The western margin of North America is a combination of subduction zone and transform fault. North of Mendocino, beneath Northern California, Oregon, Washington, and southern Canada, there is a very slow subduction zone. This subduction zone is responsible for the active volcanoes: Mt. St. Helens, Mt. Ranier, Mt. Hood, Mt. Adams, and others. South of Mendocino is the San Andreas Fault, which veers offshore to the Gorda Rise, which is a spreading center.

The dotted line along the Oregon/Washington coast is the subduction zone. Contrary to the previous statement about transform faults always being at right angles to the spreading center, the transform fault that trends east-west from the southern end of the Gorda Ridge is not at a right angle. Notice also that there are many quakes to the east of the Gorda Ridge, indicating that deformation is occurring in a region that, in the ideal case, is fairly quake free. It turns out that this region is deforming and rotating, so the occurrence of quakes within the plate is not surprising.

MARINE SEDIMENTS

INTRODUCTION

Marine sediments are formed by several processes. Detrital fragments of rocks and minerals can be carried to the sea from distant, upland sources. Alternately, they can be formed in place by biological or chemical processes operating either at the site or very close by. Marine sediments can be grouped into three categories, based on their composition and mode of origin.

- Terrigenous - grains which have been eroded from the land and carried to the marine environment, typically by rivers, wind, glaciers, slumping and mass wasting.

- Biogenic - Fragments derived from biologically precipitated skeletal material, usually broken down by physical and biological erosion. These are mainly calcium carbonate but a minor fraction can be siliceous.
- Authigenic - Chemical deposits which are usually formed by precipitation from solution in the environment of deposition.
- Evaporites, which are precipitated under hypersaline conditions have been included in this group.

Each sedimentary suite carries a record of both its origin and its ultimate environment of deposition. Sediment composition provides important clues about the location of origin of the individual grains, and textural characteristics provide information about energy levels during transport and deposition, especially in areas dominated by clastic sediments of terrigenous origin.

This chapter has two primary goals.

- The first is to help the reader understand how different sedimentary assemblages are formed and how to use these patterns to interpret the conditions that are responsible.
- The second is to consider how the oceanographic processes discussed in the previous chapters act on marine sediments to transport them from site to site and to distribute them as the deposits we observe in modern and ancient marine settings.

The discussion starts with textural characteristics common to all sediment types regardless of their origin. These sediment properties have been discussed in a number of textbooks dedicated solely to the subject of sedimentation. They are described only generally below, and the student seeking more detailed descriptions is referred to Friedman, et al., for an excellent treatment of the subject. Building on this, each major sediment type and the rocks they ultimately form are considered. Once the groundwork has been laid for understanding the characteristics of marine sediments, sampling and analytical methods are briefly outlined. Finally, the mechanisms most important in transporting sediments in modern marine systems are considered.

SEDIMENT CHARACTER

Regardless of the source of a particular sediment, certain physical characteristics are important in describing what the sediment "looks like." Primary among these are size, sorting, shape and color. These sediment properties have been discussed in a number of textbooks dedicated solely to the subject of sedimentation. They are described only generally below, and the student seeking more detailed descriptions is referred to Friedman, et al. for an excellent treatment of the subject.

Measuring Sediment Character

The physical measurement and statistical treatment of grain-size data are basic tools in the investigation of marine sediments. Characterization of the size population in a sediment sample can be either directly measured physical dimensions or their "hydraulic equivalents," which are based on the settling velocities of quartz spheres. The following describes the most common methods of grain-size determination.

Sieve Analyses

Perhaps the oldest, but still widely accepted, method of grain-size determination uses a nested set of sieves in which the size of the mesh is progressively smaller down the stack. The screen of each sieve is woven from brass or stainless steel wire to form square openings. Because of this geometry, the width of the intermediate axis is the critical determinant of grain size in sieved sediments. The sieves are agitated either mechanically or by hand, and each size class is trapped on the mesh that is too small for it to pass through. A nest of screens will, therefore, separate the sediment sample into groups of grains that range in size between the larger sieve through which they just passed and the sieve on which they were caught.

The weight of each group is expressed as a percent of the total sample weight. sieve analysis guide The accuracy of size determination is a function of several factors, each of which can introduce errors if not carefully considered. With repeated use, the wires that make up the screen may stretch or move out of place so that the dimensions of the openings are no longer true. Touching the finer screens or forcing grains through the mesh at any size are among the most common causes of damage. Another problem is that, grains tend to get stuck in the mesh. When the next batch is run, some grains are freed from the prior batch and new ones are caught. An accurate sample characterization will occur only when some sort of equilibration takes place - for each grain caught, one is released.

Thus, careful and consistent handling of the sieves is important. Finally, the amount of sample that is put through the sieves can affect the accuracy of the analysis. If too little sand is used, the small errors become more significant. When the sample is too large, one of the screens in the nest may become overloaded and not all the particles can pass through. A sample of 75-125 grams is ideal for an 8-inch diameter screen set. While all this seems fraught with peril, experience has shown that careful and consistent handling of samples can lead to very acceptable results. Sample analysis with sieves is relatively easy and quick, and the results can be both reliable and repeatable when size classes of quarter- or half-phi units are used. The mathematical transformation of raw sieve data to mean and sorting is discussed in a later section.

Pipette Analyses

The grain-size distribution of muddy sediments is commonly determined by pipette analysis. The sample is placed in a one-liter graduated cylinder filled with distilled water and a dispersing agent to prevent clay flocculation. The mixture is agitated, and 20-ml aliquots of sediment-laden water are taken with a narrow suction tube at specified time intervals and depths. The water from the pipette is evaporated in 50-ml beakers and the weights of the containers with and without sediment are used to determine the amount of sediment contained in each aliquot. Published settling rates are used to determine of the percent of sample in each size class. The technique is described in detail in Folk's 1974 laboratory manual. pipette analysis guide. Successful pipette analysis requires a very accurate balance and infinite patience.

Accurately determining the weight of sediment suspended in a 20-ml aliquot placed in a 50-ml beaker might be likened to weighing an elephant with and without a fly on its back. The weight of contaminants seemingly as minor as fingerprints on the glassware can have a discouragingly large impact on accuracy. Filtering the aliquot and weighing the dried sample reduces the problem of weight differential, as long as glass-fiber filters are used to minimize changes in filter weight under varying temperature and moisture conditions. Despite all of this, pipette analysis remains as a widely used and practical technique for the analysis of muddy sediments.

Rapid Sediment Analysers

Some of the potential drawbacks to sieve and pipette analysis can be overcome with careful sample handling, but some cannot. Sieve analysis does not take into account either the density or the shape of individual grains. Also, the analysis of large numbers of samples can be a tedious process. Finally, the mesh size of sieves cannot be made small enough to adequately divide muddy sediments into enough size classes for accurate statistical treatment. Rapid sediment analyzers were designed to address the first two of these problems.

The most popular RSA is a settling tube - a vertical cylinder one to two meters high filled with distilled water. A pre-weighed sediment sample is introduced at the top of the water column and is allowed to settle. At some point near the bottom of the tube, the sediment is collected on a pan that is connected to a very sensitive balance, and the weight of accumulated sediment is recorded over time. Either empirical or theoretical relationships between grain size and settling velocity are used to assign appropriate size-weight classes. In earlier systems, time vs. weight data were manually recorded and converted to size classes. Cost-effective, computerized systems are now available and are an integral part of most RSA's. Some RSA's use a pressure transducer that is placed at the measurement point near the base of the tube. As particles fall below the transducer, the pressure of the overlying water/ sediment mixture is reduced, and these changes are converted to sediment weight much like the hydrometer that is used to test acid densities in automobile batteries.

A major criticism of this family of RSA's is the small change in sediment/ water density relative to the level of accuracy for available transducers. Despite their increasing popularity, RSA's are not without their own set of potential problems. Among the most fundamental of these, RSA's measure sediment size as a function of fall velocity in a column of water. While a system that measures "hydraulic size" is optimal in some respects, the hydraulic behavior of grains falling through a fluid column versus grains bouncing along a bed involves somewhat different sets of forces. Also, RSA's are typically calibrated with quartz or glass spheres; populations are expressed in terms of these "equivalent diameters" and are, therefore, not a measure of actual grain size in projected cross section.

Comparison to sieve sizes can, therefore, be sometimes misleading. A second set of problems is related to grain interference. Turbulence during grain release can interfere with normal settling patterns. In addition, groups of spheres settle at different velocities than single spheres of the same size; thus, fall depends in part on the number of spheres in a group. Gibbs, And finally, the walls of a settling tube can interfere with natural settling. Most of these problems can be solved with careful design and calibration.

Other Optical Means of Determination

Because of slow settling rates, muddy sediments cannot be analyzed with a standard RSA. Several optical techniques have been developed to solve some of these problems. They are much faster than the older pipette analyses discussed above. The Coulter counter, originally designed to count blood cells, has been used to count and size individual grains. Another family of instruments measures grain concentration using one or more lasers.

Like the settling tube, the resulting size classes are based on known settling rates and the loss of progressively smaller grains over time. In all cases, the sample must be thinned to avoid the confusion that might be caused by multiple grains in proximity to one another. In addition, calibration is still a subject of active debate. While some researchers consider laser particle counters to still be in a developmental stage, they may still provide the only option when more traditional methods are impractical.

Mean-Grain Size

Sediment grains occur in a wide range of sizes from microns to centimeters. Grain size is usually expressed as a projected cross section, with the assumption that the particle is roughly circular. Wentworth divided sediments into four size categories based on grain diameter: cobble/boulder, gravel, sand and mud. Mud can be further divided into silt and clay. Grain size is expressed in millimeters, and the size distribution of clastic particles in naturally occurring sediments is basically logarithmically distributed. normal probability guide Because of the logarithmic distribution, Krumbein proposed an alternate system, the phi scale, in which: $ø = -\log_2$. An unfortunate aspect to this relationship is that as grain size gets coarser, phi size gets smaller. Most researchers simply treat phi statistics as a computational necessity and convert the answers to millimeters. For those of you who choose to work in phi units, the earth is -33ø.

Shape

Particle shape describes the three-dimensional character of a grain and may be expressed in several ways. It may be referred to specific end members. Along the same line, sphericity is a numerical measure of how closely the grain approaches a sphere. In siliclastic sediments, individual grains tend to be more equidimensional; especially if they have traveled long distances and have undergone significant abrasion. In contrast, on beaches where sediment is derived from cliff erosion grain shape might be widely varied and the tendency for gravel and cobble-sized clasts to move either landward or seaward is strongly affected by their shape.

Grain shape is perhaps most important in carbonate sediments where skeletal characteristics of the parent skeletal material has not yet been altered by abrasion. Platy grains are often more easily eroded and will settle more slowly than round grains of similar size. A final characterization of shape is angularity. This is a qualitative measure of the curvature of the corners of a particle independent of sphericity and it reflects the amount of abrasion the grain has undergone. The degree to which a sedimentary particle will be rounded depends on

- the energy level of the transporting process,
- the duration of that process and
- the durability of the original grain.

Mathematical Determination of Grain Size

Standard techniques have been developed to quantitatively compute and display the results of a grain-size analysis. In the method based on sieve analysis, the weight of the sediment caught on each sieve in the stack is tabulated and divided by the total weight of the sample to compute its weight percent. The cumulative frequency for that size is computed by adding the weights for sediment caught on that sieve and all those higher in the stack. A general idea of the sediment character can be drawn from a histogram that graphically shows the relative abundance of each size class. Better-sorted samples will have a single peak, and the abundance within adjacent size classes will drop off sharply. Poorly sorted samples will have a less-distinct mode or even multiple modes, the latter being particularly common in carbonate sediments.

Computation of the sample mean utilizes a statistical measure called the "method of moments." To visualize this, imagine a playground seesaw that has been divided into equally spaced intervals with each successive interval labeled in order to correspond to the sieve sizes in the stack. If the sediment from each sieve is placed in the center of the corresponding interval, it would be possible to determine the mean grain size by sliding the seesaw back and forth to find the point at which it balances. The method of moments does this mathematically using statistical relationships that take into account widely accepted statistical principles. The logarithmic distribution of sediments in nature is taken into account by using the phi distribution discussed above. This technique is discussed in Friedman, et al.. method of moments from Friedman, et al..

A less accurate but much simpler method for calculating grain distribution statistics uses only a few points from the plot of grain size versus cumulative frequency. The critical points on the resulting curve include the 50th percentile, the 16th and 84th percentiles which theoretically represent one standard deviation away from the mean in a normal population. The 5th and 95th percentiles can be used with the other three points. The sample mean is computed by averaging the median and the values one standard deviation unit to either side:

$$(Mz) @ \frac{i_{16} \cdot i_{50} \cdot i_{84}}{3}$$

Folk recommends that standard deviation be computed using information on the sample within two standard deviation units of the mean. This is particularly useful in carbonate sediment analysis where the "tails" of the size distribution curve can be quite variable and, therefore, contain important information. His formula for inclusive standard deviation is:

$$(v) @ \frac{i_{84}\ 0\ i_{16}}{2}.$$

If the sample is distributed such that values for the 5 and 95 cannot be derived from the graph, then a less accurate determination can be made using:

$$(v) @ \frac{i_{84}\ 0\ i_{16}}{4} \cdot \frac{i_{95}\ 0\ i_{5}}{6.6}$$

Discussion of other statistical parameters may be found in Friedman, et al.. Papers by Folk and Moiola and Weiser discuss the methods of analysis and statistical handling of the results.

Displaying Sediment Data

Krumbein introduced the idea of using a ternary diagram to graphically display sediment texture by plotting the relative abundance of sand, silt and clay. Since then, various authors have refined this plot by assigning boundaries among different sediment types. In most classifications, the end members sand, silt and clay are considered to be reached when more than 75 percent of the sample is within that size range. Mixtures of sand, silt and clay are named by the dominant member with a modifier to acknowledge grains of lesser importance.

Other workers have defined classification boundaries using natural breaks that occur in sediment types at a particular site, arguing that this system takes into account identifiable gaps in the sedimentary continuum that may reflect important shifts in the controlling processes. The obvious problem with this approach is the resulting difficulty in directly comparing sediments from one area to another using only nomenclature. Using this information, it is possible to map textural facies, which show the distribution of sediment types in a particular area.

In addition to the ternary classification just discussed, maps can be constructed to show mean grain size, median size, coarse/fine sediment ratios or other sediment distributions that somehow reflect gradients or breaks in physical-process controls. To add stratigraphic context, some investigators have converted sedimentary suites to equivalent facies. The implication is that a surficial pattern of sediment types will also be reflected in a vertical column such that adjacent units are stacked one on top of the other.

To qualify as a facies, a lithic unit must:

- Consist of a set of mappable characteristics,
- Occupy a restricted part of a stratigraphic unit, Moore, and
- Be definable based on purely descriptive parameters

In this concept of facies, introduced by Gressly, a sediment facies is a mappable unit that can be distinguished from other deposits based on grain size and sorting, grain type, carbonate content or some other textural characteristics. The intent of a facies map is to show some gradient or pattern that reflects changes in present or past conditions. A particular suite of sedimentary textures and structures reflects the processes that occur in that environment. Together with data on flora and fauna, these sedimentary facies can be used to define an environmental facies. When using such mixed terminology, however, care should be taken to ensure that a facies name conveys a real sense of change or a contrast between differing conditions of deposition.

THE SIGNIFICANCE OF TEXTURAL CHARACTER

For the purpose of the following discussion, we have divided clastic marine sediments into two broad categories: terrigenous and carbonate. It is important to understand that the processes that are responsible for their formation and deposition can be profoundly different Terrigenous sediments are most often of siliciclastic origin. In general, they are derived from the erosion of upland or coastal areas and are transported over considerable distances to their ultimate environment of deposition.

Terrigenous sediments can also be eroded carbonate that has been lithified and uplifted. In this instance, however, the processes of erosion, delivery and final deposition are generally similar to those of siliciclastic sediments. In contrast, first-generation carbonate sediments were by and large produced by living organisms in or near to the area where they ultimately came to rest. While this differentiation is somewhat oversimplified, it is still useful for considering the different forces that control the distribution of sediments from each of these two groups. The downplaying of a third group of sediments - hypersaline evaporites and authigenic minerals - is not intended to imply any lack of importance. Rather, it simply reflects their more restricted distribution in nature and a narrower treatment in the published literature. For terrigenous sediments, a reasonably clear relationship exists between energy level and grain size. Coarser sediments generally reflect higher energy, while quieter areas are dominated by mud.

On land, progressively finer sediments are deposited as the slope decreases and the rivers slow down. In the marine environment, energy levels and, therefore, sediment size are controlled by such factors as wave action, exposure, tidal range and water depth. In carbonate environments, however, this relationship is less reliable due to the effect of the in-situ biological origin of sediments. The size of carbonate grains is also strongly controlled by the skeletal architecture of the organisms from which the grains are derived. This strong biological overprint does not mean that physical processes are any less important in carbonate depositional systems.

To the contrary, physical factors still impart a strong signature on carbonate grains, but one that is made subtler by biological effects that must also be taken into account. What is required for successful paleoenvironmental reconstruction from carbonate sediments is a careful examination of both textural parameters and the constituents that make up the deposit. Sorting is related to uniformity of both the energy regime and sediment supply. For example, wave action is usually the most important controlling factor along open beaches, and sand is moderately sorted, reflecting even energy and supply. In a similar environment, but with nearby alluvial cliffs, beach sediments might be poorly sorted if a significant portion of the material being eroded cannot be moved by ambient processes; the character of the source is overwhelming the ability of physical energy to sort it and. lag gravel and sand are formed.

The interpretation of sorting in carbonate sediments can be problematic. The variable grain shapes and the high porosity of some particles can make them more susceptible to transport than terrigenous grains of a similar size.

Carbonate deposits often have a polymodal size distribution that is related partly to hydraulic equilibrium and partly to the breakdown characteristics of the skeletal organisms that have lived and died within a particular setting. For example, sediment in the backreef may be a mixture of skeletal sand produced by the disarticulation of organisms that live there under normal conditions, gravel or cobbles washed in from the adjacent reef crest during storms, and carbonate mud produced by the biological breakdown of both. The mud and fine sands are in hydraulic equilibrium with quiet conditions at the site most of the time, while the coarser grains are either produced in situ as skeletal debris or are dumped behind the reef during storms. None of these can be removed by weak ambient currents and the resulting sedimentary deposit reflects a myriad of physical and biological agents. While the mixture of grain types makes interpretation of absolute energy level difficult at best, the unique assemblage of grain sizes and types make it difficult to misinterpret the environment of deposition, especially if the deposits from the reef in front and the lagoon behind can be identified.

Whatever the origin of a particular sediment and however mixed the processes have been that brought it to its final resting place, sediment composition and texture will hold important clues to those controlling factors. Whether siliciclastic or carbonate in origin, the physics of transport and deposition are the same. To understand the factors that have contributed to the formation of a particular sedimentary suite, one must become intimate with the complex interplay among sediment composition, sediment texture, and the physical and biological processes related to transport and deposition. Each of these is discussed in greater detail below.

TERRIGENOUS SEDIMENTS AND ROCKS

Sources

Most terrigenous sediments are ultimately derived from the breakdown of crystalline igneous rocks. Although sedimentary rocks are also eroded to provide grains, their constituents are generally derived from the prior erosion of igneous or metamorphic rocks. Notable exceptions include cherts, carbonates and volcaniclastic grains. The principal source of terrigenous marine sediments is river discharge.

Rivers transport more than 18 $\times 10^9$ metric tons of suspended solids to the world's oceans annually. Holeman, This amounts to a 100-km high column of sediment occupying the area of two football fields. More than one-third of this is carried by about ten rivers, with the Ganges and Amazon Rivers alone carrying 20 percent of the worldwide total. Sand and gravel rolling or bouncing along the bed are mainly deposited in estuaries and on beaches or offshore bars. The muds that travel in suspension are deposited in estuaries and coastal wetlands, largely due to a process called flocculation. As sediment is delivered to the open ocean, most of it is trapped by wave action in the nearshore environment and moves parallel to the coast, with less than ten percent of modern river suspension sediment load ever reaching the deep ocean. Friedman, et al., During major storms, however, offshore-flowing currents can reach a magnitude sufficient to move clastic sediments through the energy barrier of the surf zone and away from the shore. In some cases, these form into shore-parallel sand bars and migrate back onto the beach. In other instances, sediment is lost to the shelf until the next sea-level drop.

Terrigenous Sand

The assortment of minerals occurring in a sediment is the key to identifying the site from which it was derived. While a tedious task, identifying all the mineralogical components and their likely sources is necessary if a study of provenance is to be quantitatively useful. Unfortunately, the work expands rapidly as more sophisticated techniques are employed. Blatt, et al. point out that the typical approach is a necessary compromise between the need for detailed analysis and the amount of time available. Although hundreds of minerals have been identified, only a few are considered to be common rock-forming minerals. Most of the common minerals in terrigenous sediments are silicates, which are made up of silicon and oxygen with strong atomic bonds. The common silicate minerals include quartz, feldspars, micas, pyroxenes, amphiboles, micas and olivine. Except for quartz, these minerals have similar chemical composition of calcium; sodium, potassium and iron bound in a silicate tetrahedron.

Techniques for identification of mineral and rock grains in a sediment sample range from hand lens examination to the use of petrographic microscopes and Scanning Electron Microprobes. Identification of terrigenous rock forming minerals is not easy, but some general characteristics such as color, hardness and cleavage help in the identification. Once reduced to grains composed of individual minerals, the durability of each mineral will dictate its stability. The composition of sediment near to its original source will largely reflect the mineralogical composition of the parent rock. Even at its source, however, the sedimentary suite will also reflect the erodability of the underlying rocks under the influence of local processes. In high-latitude glacial climates, processes such as frost cracking and glacial scour may produce a different sediment assemblage when compared to the same rock type exposed to arid desert conditions, including intense heat and sand-blasting by wind-blown sand.

The further away from the source, temporally or spatially, the more important the more durable minerals become. As an illustration of this process, consider that the "average" sand or sandstone contains about 65 percent quartz with a mean size of 2? and less than 15 percent feldspar. In contrast, the igneous rocks from which they are derived contained more than 60 percent feldspar.

The disparity in composition is a reflection of the greater resistance of quartz to physical and chemical weathering in a surface environment. Feldspar is less resistant and is rapidly converted to silt and clay.

Thus, the relative abundance of quartz versus feldspars and other, less-stable minerals is a reflection of the intensity and duration of the weathering processes. Quartz, orthoclase feldspar and the micas will dominate near the source. Over time, quartz becomes the most common sedimentary mineral because of its high stability to physical and chemical weathering. Feldspars are found in areas of rapid deposition or proximal to the source area.

Terrigenous Muds

Over time, less durable minerals will break down to form silt and clay. Silts are comprised of mainly quartz, feldspar, and carbonate grains that fall into the size range of 4 to 9 ?. Friedman, et al., The term clay refers to a specific group of minerals with a basic silicate structure and composition which is combined with metallic ions to form the different clay minerals.

The particles are all smaller than 9 phi and can be seen only with an electron microscope. clay minerals The composition is similar to the micas and other silicate minerals, but the structure is unique in having two different types of layers combined in different patterns with layers of water molecules.

Classification of Terrigenous Rocks

The underlying sedimentary character is the cornerstone for classifying terrigenous rocks, with the grain size of sediments being generally translated into a corresponding rock type. Claystones, siltstones, shales, sandstones and conglomerates represent rocks made up of increasingly larger grains. Because the classification of a particular rock is in part related to the distribution of its composite grains, conglomerates and sandstones can be further classified on the basis of mineralogical components. Most schemes use a ternary diagram with quartz, feldspar and metamorphic/extrusive rock fragments as the three end members to develop a classification. These end members are plotted by percentage to determine the basic rock type. On the ternary diagram, arenites are those rocks with sand-sized grains, predominantly composed of quartz. Arkosic rocks contain much higher proportions of feldspar and are often poorly sorted. Graywackes are usually darker rocks comprised with a greater percentage of angular rock fragments. The fine-grained sediments and rocks may be divided on the basis of the dominant mineral composition. Accessory minerals such as glauconite or pyrite are often used as modifiers in the rock name, *i.e* glauconitic quartz sandstone.

CARBONATE SEDIMENTS AND ROCKS

Modern carbonate sediments are comprised of three principal minerals: aragonite, calcite and dolomite. The mineral type is determined by the organism from which the grains are derived. Once lithified, these will generally form limestone, although chemical diagenesis or lithification in specialized conditions that concentrate magnesium can form dolomite.

Sedimentary Constituents

Carbonate grains can be broadly divided into two types. Skeletal grains have a biochemical origin, and are formed as internal or external skeletal units of plants or animals. Non-skeletal grains are formed by a variety of physical and chemical processes. The following describes the most important constituents of carbonate sediments and provides some very basic keys for their identification. While most of the identification criteria are easily seen in hand sample, a few require examination in thin section.

Biogenic Grains

Corals are among the more important carbonate producers in tropical seas. The dominance of coral fragments in a sample is an important first step in recognizing the presence of a nearby reef. Coral fragments in hand specimens are massive and blocky with a whitish opaque-to-translucent surface. When subjected to constant abrasion, they will take on a high polish. In thin section, they have a blocky texture and are transparent; internal structure related to the original skeleton may be apparent. Coralline algae from the family Corallinaceae are likewise indicators of a nearby hard substrate or a reef. Red algae are divisible into:

- Crustose corallines which develop sheet-like crusts, blunt knobs or branched protuberances and
- Lithothamnion), and articulate corallines whose members are small, erect and usually segmented plants.

Many fresh coralline grains will retain the pale, pinkish color of the parent plant. This is not a reliable feature, however, as it quickly fades. The surface of the grain will usually have a "sugary" texture but can be easily mistaken for coral, especially in polished or less-than-pristine samples. In thin-section, however, corallines possess striking laminae and lattice-like patterns that are difficult to miss, even at low magnification. Molluscs are important constituents of virtually all carbonate environments as whole shells or fragments. In non-reefal areas, molluscs may be the dominant contributors to the sedimentary record.

They are important contributors to both lagoonal sediments and those found along the shelf in front of the reef. At the phylum level, molluscs are of limited usefulness for environmental interpretation because of their widespread occurrence and variable mineralogy. However, identification to genus or species level can greatly increase the resolution of environmental discrimination. In hand samples, identification is facilitated by the curvature of the parent shell.

Also, molluscan material retains the highest natural polish of any carbonate grain. In thin section, the composite nature of the shell is reflected in a "herringbone pattern." Foraminifera are one-celled animals that usually secrete calcareous tests; in essence, they are amoebas with shells. Forams are commonly less than a millimeter in diameter, but larger individuals do occur; members of the genus Leptocyclina with lengths over 10 cm occur the Oligocene of Puerto Rico. Benthic forams are more common in shallow, sandy areas, and certain species can be tied to specific environments. Their thicker tests adapt them to life in shifting sands, in contrast to planktonic varieties that have thinner shells to facilitate floating in the water column. Their wide distribution limits the usefulness of planktonic forams for paleoenvironmental determination. Specialized groups of encrusting forams are found on the surface of hardgrounds, and are particularly abundant on dead reef substrates. Without careful examination, small bits of forams and gastropods may be confused with one another, either in hand specimen or thin section. The key to discriminating between the two is the segmentation of the foram body cavity into chambers.

Gastropods contain a single, coiled shell cavity, while forams have numerous and separate chambers. Halimeda are common inhabitants of sandy or grassy bottoms, but can also occur along hardgrounds and reefs to depths of over 200 meters. The plates break down into readily identifiable, flakelike grains that have a chalky surface with small pores. In cross section, the grains appear as a sandwich of denser material surrounding a porous cortex. In thin section, the meshwork of channels in the central section is even more obvious and is manifested in a meshwork of tubes and circles. Echinoid fragments are common in sandy and grassy areas. Body segments are platelike and can be easily mistaken for coralline algae or Halimeda fragments unless surface ornamentation is present. In contrast, echinoid spines are distinctive in their character.

They are 1-2 mm across and usually have ribs that run lengthwise along the spine. These ribs and the shiny surface of the spines are useful in separating them from similarly shaped, articulate-algal segments. In thin section, echinoderm fragments are easily identified by their characteristic unit extinction. A number of other grain types occur in limited numbers in many carbonate sediments. Sponge and gorgonian spicules are distinct in their shape and can be recognized on the basis of appearance. Because they are very small, they are easily swept away from the moderate- to high-energy environments that both animals inhabit.

Preservable hard parts of worms and the calcareous tubes that they secrete are likewise broken down and rare. Ostracods, members of the Crustacea, have bivalve shells, which can be highly ornamented with spines and nodes. They are usually a millimeter or less in maximum diameter, although a few may be as large as a centimeter. A number of species may be recognized, and since the whole shell is often preserved, ostracods can be used in the same manner as foraminifera for environmental determination.

Non-Skeletal Grains

Ooids are spherical grains formed by accretion about a central nucleus. There is still disagreement over the degree to which formation is the result of biochemical versus inorganic physio-chemical precipitation. Whatever mechanism is responsible, the concentric layers of these grains represent successive episodes of accretion as the grains roll back and forth across the bottom. The size to which an ooid may grow is limited by the energy regime of the environment of deposition. Friedman et al., Constant agitation encourages the successive addition of new carbonate, much like the creation of elements of a snowman. Peliods are small, ellipsoidal or spheroidal grains, which lack distinct internal structure. These result from several processes. Very fine-grained, unconsolidated carbonate sediments can aggregate and become cemented, perhaps aided by bacterial precipitation of aragonite. Many peloids are fecal pellets produced by sediment-ingesting organisms or plankton feeders.

These consist of clay-, silt- or sand-sized grains loosely bound by organic mucus. Because they are friable and easily disaggregated, cementation or rapid burial is necessary for their preservation in turbulent areas. From a sedimentary standpoint, peloids are important because they provide a mechanism to enhance deposition of muddy sediments. Composite grains are comprised of multiple grain types. They may be formed by erosion of pre-existing rocks and cemented sediments or by the aggregation of loose grains into grapestone. Along the Bahama Banks, the latter occur in areas of intermittent agitation. Imbrie & Purdy, Along adjacent and more-energetic bank margins, ooids are the norm.

Mud is an important constituent of many carbonate deposits. It may be produced by the direct precipitation of calcite or aragonite from solution, comminution from coarser carbonate material through bioerosion, or the secretion of fine aragonite needles within the organic tissue of algae such as Penicillus or Udota. The presence of carbonate mud implies an inability of the ambient processes to remove fine-grained sediment that has been produced locally. Many muddy carbonate sediments will also have a significantly coarser fraction, derived from organisms living in the area. This pattern is hard to reconcile with clastic-based models in which mixed sediment sizes are less common. This type of situation is one important cause of the polymodal grain-size distributions.

Carbonate Rock Classification

In general, but certainly not always, carbonate rocks are formed by the cementation of carbonate sediments after burial and/or uplift. In 1962, two classifications were proposed that take into account both the origin of the constituent grains and the processes associated with their deposition and cementation. Both are still in widespread use today. Folk's classification focuses on the petrographic analysis of the larger grains and the character of the matrix. The matrix can be finer grains bound together, or a diagenetic replacement of an earlier matrix element. Rock names are comprised of three parts.

The prefix is derived from the dominant types of larger grains that are present:

- Bioclastic debris, ooids, intraclasts and pellets. The middle part of the name is based on the matrix size. A matrix with grains larger than 10 μm is termed sparry and finer mud matrix is called micrite. As a rough rule of thumb, if the crystals are easily visible at 40X magnification, they are probably sparry. The last part of the name is based on the size of the allochems. Gravel-sized allochems are identified by the suffix -rudite; sandy allochems are termed arenites.

A contemporaneous classification by Dunham is more useful for describing hand samples and uses the percent of mud in the rock as a proxy to energy regime and sediment supply. Increasing quantities of mud in the matrix can signal either a drop in energy level or an increase in the supply of fine-grained material. It is a textural scheme based primarily on whether the sand-sized and larger grains touch one another or are "floating" in a matrix of mud or replacement cement. Grain-supported rocks are divided into grainstones, which have no mud in their interstices, and packstones that do. Mud-supported rocks with more than 10% of their volume occupied by larger grains are classified as wackestones. Those with greater than 90% matrix are termed mudstones.

The inference is that grain-supported rocks were formed from sediments in well-washed, higher-energy areas, whereas other rock types with more mud reflect progressively calmer conditions. A special class, boundstones, includes rocks of biogenic origin that were bound together by encrusting organisms. Embry and Klovan later added several other rock types to this scheme, such as rudstones, framestones. More recently, Carozzi proposed a classification that integrates many of the ideas of both schemes. The relative abundance of larger grains is more finely subdivided than in Dunham's classification and the size descriptors of Folk are maintained. However, the classification is a bit clumsy in that the descriptors can be as long as the sentence it was designed to replace.

Each classification has its own inherent strengths and weaknesses. A student's ultimate choice is typically dictated by which scheme best suits specific emphasis or interests. In actual practice, many researchers borrow from several schemes and liberally add modifiers for "classifications."

EVAPORITES AND AUTHIGENIC SEDIMENTS

Evaporite deposits can comprise a significant proportion of ancient rocks, but seem to be less well represented in modern marine environments. Modern evaporites form in sabkhas, salinas and interdune environments, which are all in subaerial, or shallow subaqueous environments where evaporation rates are high. Three evaporitic deposits are common in nature: halite, anhydrite, and gypsum. Halite forms in an evaporative basin in conditions that are more restricted than for the formation of gypsum. However, experimental results show that gypsum and halite coprecipitate during intermediate stages of evaporation, explaining the common occurrence of calcium sulfate minerals in halite beds in ancient rocks.

Boggs, Subtidal evaporite environments have also been identified in the geologic record and owe their origin to the injection of hypersaline waters through groundwater processes referred to as reflux. Authigenic minerals form in place in the depositional environment. The formation of pyrite is largely driven by bacterial action under the influence of sulfate reduction in dark organic rich muds under reducing conditions. In this environment, ferrous iron, hydrogen sulfide, and native sulfur react to form iron sulfides. Glauconite occurs as sand-sized green, earthy pellets found in marine sediments and as fill in foraminifera tests.

SEDIMENT TRANSPORT

Terrigenous sediments are typically introduced to the ocean by either fluvial, eolian or glacial transport. Their distribution in the marine environment is primarily a function of the interaction between the strength of waves and currents and the size of individual sediment grains. Although produced locally, carbonate sediments are also redistributed by these same processes. Sediment particles respond to hydraulic forces such as shear and lift, whose effects are related to current speed, particle size, shape and density. While quantitatively modeling sediment transport can be very difficult, it can generally be thought of as moving water exerting both lift and drag on a sediment grain at rest. As the velocity of fluid flow over the bed increases, both lift and drag increase.

At some point, the fluid exerts sufficient force to cause the grains to move. This fluid force must overcome both gravity and any friction exerted by adjacent grains. Once the force is great enough to overcome these, the grain begins to move, either along the bed or in the water column. Transport in the ocean obeys the same hydraulic laws that apply in streams, but the forces available for transport are more varied and the motions are more complex. The remainder of this chapter focuses on the effects of these processes on marine sediments. Serious theoretical sedimentologists will argue that any discussion of sediment transport that is decipherable by the average geologist is, by definition, totally inadequate inasmuch as it ignores the necessity for things in nature to be complicated beyond our practical understanding. While this cynical statement may reflect the authors' poor grasp of things theoretical, the problem nevertheless remains of expressing these ideas in a form that can be grasped conceptually by the student who is interested more in the general relationships that occur in nature than in generating rigorous predictive solutions. We, therefore, beg the forgiveness of those that prefer shear stress to velocity and dimensionless grain number over sediment size and hope that our simplistic approach is informative to the student and minimally repulsive to the rest.

Sediment grains move in three modes. Grains in suspension move with the water mass in which they are contained. This is generally fine-grained material, but larger grains can be carried if the velocity of water motion is very great. Bed-load transport involves particles that are too heavy to be put into suspension and are moved along the bottom in a rolling or sliding motion. The layer of active transport is only a few grains thick. Intermediate between the two is a process called saltation in which the particles move in a series of jumps. A particle is thrown into suspension either by fluid turbulence or grain impact, and moves with the water until it falls again to the bottom. This is an important process in the movement of sand in both nearshore and shelf settings because wave action can periodically throw sediment into the water column to be moved by weaker, unidirectional currents to a new spot. Repetition of this with each wave can result in effective grain transport.

Grain Settling

Because grains in motion spend much of their time in the water column, it is important to understand the interactions between the two. In water, the rate at which a particle settles is a function of both grain and fluid properties. In calm water, the settling rate of fine-grained sediment is adequately approximated by the Stokes Equation:

$$V \, @ \, \frac{gD^2}{18} \frac{(d_s \; 0 \; d_w)}{p_w}$$

where:

V= fall velocity in cm/sec

g = gravity

D = grain diameter in centimeters

d_s = density of the particle

d_w = density of the fluid in the same units or 1.0 for water at 20° C

μ_w = viscosity of water; 1×10^{-3} at 20° C

For gravel sized particles, the impact formula is:

$$V = \sqrt{\frac{4gD}{3} \frac{(d_s - d_w)}{d_w}}$$

Fall velocity can be plotted for these two formulae and the results compared to empirical measurement. Gibbs, et al., While the theoretical lines plot consistently above the experimental data, it can still be seen that there is a consistent relationship between smaller grain size and declining fall velocity.

Transport in the Marine Environment

Unfortunately, the water in most marine systems is neither "ideal" nor motionless. Current-flow patterns can vary greatly from one environment to another. In estuaries, currents generated by tides dominate. While unidirectional at any one time, these currents reverse with each change of the tide. Net transport in a tidal system reflects the relative strengths and durations of seaward vs. landward currents. Once we move outside the protection of the estuary, wave-induced motion is added to the effect of tidal currents and net transport is the result of various combinations of unidirectional and oscillatory flow. The orbital motion of passing waves produces oscillatory flow that decreases exponentially in its magnitude from the water surface toward the bottom. At a water depth roughly one half of the wave's length, the orbits begin to interact with the bottom, causing bed shear.

The orbital motion becomes increasingly elliptical toward the bottom until the motion is transformed into back-and-forth oscillations with more and more energy transferred to the bed as water depth shoals. The more intense flow under the crest results in greater the transport. The water depth to which sediment will be moved by waves is a function of particle size and the wave regime that exists. During fair weather, wave base is on the order of 10-15 meters. During storms, this can increase dramatically. Off the Texas coast, hurricane waves are capable of moving sediment anywhere across the continental shelf. Along the shelf edge west of Britain, the intense wave climate can stir fine sand at a depth of more than 180 meters for more than 20 percent of the year.

There is strong evidence that total amount of sediment transportation during a single storm can be much greater than the total for the rest of the year. On the north coast of St. Croix, sediment export from the reef at water depths of 15-30 meters is on the order of 6,300 kg per meter of shelf per year during normal sea conditions. Small annual storms with waves reaching 5 m in height can remove at least 4,400 kg of sediment from the same area in a matter of hours. In 1989, Hurricane Hugo removed in excess of 48,000 kg/m from these channels, an amount equal to nearly eight years of day-to-day sediment export.

Predicting Sediment Transport

Particles that are settling in a moving column of water will do so at rates that are slower than those in calmer water. Several useful diagrams have been constructed to approximate the transportability of sediments under different flow regimes. The Shields Diagram relates the initiation of sediment motion to shear developed along the bed. Unfortunately, shear stress is nearly as difficult to measure accurately as it is to explain quantitatively. The shear-stress relationship can be used to predict the velocity one meter above the bed caused by waves of various heights and lengths. If we can then determine the current velocity that is needed to initiate or sustain sediment transport, we can then relate wave character to bed stability. While somewhat dated, the Hjulstrom Diagram is still useful because current speed is more easily visualized than other factors now in vogue.

The principle that is illustrated states that a sediment grain of given size will require a certain current velocity to pluck it from the bed. The flow needed to keep that grain in suspension is slightly less. In sand and coarser sediment, grain size is directly proportional to the velocity needed to either erode grains or keep them in suspension. In sediments finer than medium to fine sand, grain-to-grain adhesion and electrostatic charges increase the resistance of the bed to erosion. Clay and silt particles may be especially difficult to erode, and compaction and cohesion lead to the formation of a smooth surface that promotes laminar flow.

It is difficult to make quantitative estimates of the threshold for moving cohesive particles of compacted fine silt or clay, but the velocities can exceed those necessary to move gravel-sized material. The best that can be said is that mud beds will be eroded at current velocities on the order of 20 to 30 cm/sec, provided the water content of the sediments exceeds 80 percent. Below that, the bed becomes increasingly resistant to erosion. Whether we use the Hjulstrom Diagram to illustrate the general principle or the Shields Diagram to make precise predictions, the message to take from all of this is that sediment transport is a predictable phenomenon that follows well-defined physical laws and at least conceptually operates in a predictable manner. We can say that medium sand will be suspended by currents on the order of 20-25 cm/sec; it will stay in suspension until flow drops below 15-18 cm/sec. In a wave dominated environment like the shoreface at a depth of 10 meters, sand suspension can be initiated by waves only one meter high with a period of 4-5 seconds. From a general geological perspective, this level of accuracy is quite adequate.

SEDIMENT-ORGANISM INTERACTIONS

The transport regime of a particular environment is a function of the physical, biological and gravitational processes that operate in that area. In many instances, sediment transport is controlled by a combination of these factors. For example, sediment on steep slopes can be moved by currents too weak to move the same sediment on a flat surface. Plants can inhibit the transportation of both carbonate and terrigenous sediments by stabilizing the bottom in much the same way as vegetation does on land. Seagrasses, sponges, mangroves, and algal mats all serve to reduce both wave and current energy at the bed, thus stabilizing fine sediments. In the Bight of Abaco, Neumann, et al. found that a five-fold increase in currents was required to erode sediment covered by mats of blue-green algae when compared to threshold velocities for the same sediment devoid of algal cover. Many marine animals obtain their food by filtering suspended solids from the water. These include mollusks, barnacles and copepods. In addition, several animals ingest sediment that is passed through their guts and excreted as pellets. Pelletization effectively increases grain size by aggregation, thereby lowering the potential for sediment transport.

Thus, grains with the transport characteristics of sand can occur in an area where only mud-sized particles are being produced. Burrowing in marine sediments by a host of vertebrates and invertebrates actively mixes sediments, and in concert with other physical or gravitational processes, can accentuate physical transport. Resin casts made by pouring epoxy into the burrows of the shrimp Callianassa have revealed that bioturbation by these shrimp can extend more than one meter into the bottom. Excavated sediment is thrown into the water column by currents created the shrimp and settles back to the bottom in a conical deposit - a process that has been likened to a small underwater volcano.

On St. Croix, Roberts, et al. calculated that sediment suspended by this process would be moved by weak but unidirectional currents at an average rate of nearly 100 kg/m-yr in currents of only 5-10 cm/sec. In effect, the shrimp does the work of sediment suspension while the background current provided the impetus for incremental transport. In estuaries, burrowing clams and crabs can turn over tremendous quantities of sediment, destroying diagnostic sedimentary structures in the process. In some instances, this process can sort bottom sediments, leaving those at the surface more susceptible to transport. The shrimp burrowing described above also removes organic material from the sediment, thereby making it more susceptible to transport.

TEMPERATURE, SALINITY, AND DENSITY IN PHYSICAL OCEANOGRAPHY

AN OVERVIEW

Heat fluxes, evaporation, rain, river in flow, and freezing and melting of sea ice all influence the distribution of temperature and salinity at the ocean's surface. Changes in temperature and salinity can increase or decrease the density of water at the surface, which can lead to convection. If water from the surface sinks into the deeper ocean, it retains a distinctive relationship between temperature and salinity which helps oceanographers track the movement of deep water. In addition, temperature, salinity, and pressure are used to calculate density. The distribution of density inside the ocean is directly related to the distribution of horizontal pressure gradients and ocean currents. For all these reasons, we need to know the distribution of temperature, salinity, and density in the ocean. Before discussing the distribution of temperature and salinity, let's first define what we mean by the terms, especially salinity.

Definition of Salinity

At the simplest level, salinity is the total amount of dissolved material in grams in one kilogram of sea water. Thus salinity is a dimensionless quantity. It has no units. The variability of dissolved salt is very small, and we must be very careful to define salinity in ways that are accurate and practical. To better understand the need for accuracy. Notice that the range of salinity for most of the ocean's water is from 34.60 to 34.80 parts per thousand, which is 200 parts per million. The variability in the deep North Pacific is even smaller, about 20 parts per million. If we want to classify water with different salinity, we need definitions and instruments accurate to about one part per million. Notice that the range of temperature is much larger, about 1°C, and temperature is easier to measure. Writing a practical definition of salinity that has useful accuracy is difficult and various definitions have been used.

Originally salinity was defined to be the "Total amount of dissolved material in grams in one kilogram of sea water." This is not useful because the dissolved material is almost impossible to measure in practice. For example, how do we measure volatile material like gasses? Nor can we evaporate sea-water to dryness because chlorides are lost in the last stages of drying.

A More Complete Definition

To avoid these difficulties, the International Council for the Exploration of the Sea set up a commission in 1889 which recommended that salinity be defined as the "Total amount of solid materials in grams dissolved in one kilogram of sea water when all the carbonate has been converted to oxide, the bromine and iodine replaced by chlorine and all organic matter completely oxidized." The definition was published in 1902. This is useful but difficult to use routinely.

Salinity Based on Chlorinity

Because the above definition was difficult to implement in practice, because salinity is directly proportional to the amount of chlorine in sea water, and because chlorine can be measured accurately by a simple chemical analysis, salinity S was redefined using chlorinity:

$$S = 0.03 + 1.805Cl$$

where chlorinity Cl is defined as "the mass of silver required to precipitate completely the halogens in 0.328 523 4kg of the sea-water sample." As more and more accurate measurements were made, turned out to be too inaccurate.

In 1964 UNESCO and other international organizations appointed a Joint Panel on Oceanographic Tables and Standards to produce a more accurate definition. The Joint Panel recommended in 1966 that salinity and chlorinity be related using:

$$S = 1.80655\ Cl$$

This is the same as for S = 35.

Salinity Based on Conductivity

At the same time was adopted, oceanographers had began using conductivity meters to measure salinity. The meters were very precise and relatively easy to use compared with the chemical techniques used to measure chlorinity. As a result, the Joint Panel also recommended that salinity be related to conductivity of sea water using:

$$S = -0.08996 + 28.2929729\ R_{15} + 12.80832\ R_{15}^2 - 10.67869\ R_{15}^3 + 5.98624\ R_{15}^4 - 1.32311\ R_{15}^5$$

$$R_{15} = C(s,15,0)/C(35,15,0)$$

where C (S, 15, 0) is the conductivity of the sea-water sample at 15°C and atmospheric pressure, having a salinity S derived from and C (35, 15, 0) is the conductivity of standard "Copenhagen" sea water. Millero (1996) points out that is not a new definition of salinity, it merely gives chlorinity as a function of conductivity of seawater relative to standard seawater.

Practical Salinity Scale of 1978

By the early 1970s, accurate conductivity meters could be deployed from ships to measure conductivity at depth. The need to reevaluate the salinity scale led the Joint Panel to recommend in 1978 that salinity be defined using only conductivity, breaking the link with chlorinity. All water samples with the same conductivity ratio have the same salinity.

The Practical Salinity Scale of 1978 is now the official definition:

$$S = 0.0080 - 0.1692\ K_{15}^{1/2} + 25.3851\ K_{15} + 14.0941\ K_{15}^{3/2} - 7.0261\ K_{15}^2 + 2.7081\ K_{15}^{5/2}$$

$$K_{15} = C\ (S,15,0)/\ C\ (KCl,15,0)$$

$$2\ f\ S\ f\ 42$$

where C (S, 15, 0) is the conductivity of the sea-water sample at a temperature of 15°C and standard atmospheric pressure, and C (KCl, t, 0) is the conductivity of the standard potassium chloride (KCl) solution at a temperature of 15°C and standard atmospheric pressure. The standard KCl solution contains a mass of 32.435 6 grams of KCl in a mass of 1.000 000kg of solution. Lewis (1980) gives additional equations for calculating salinity at other temperatures:

$$S = 0.0080 - 0.1692\ R_t^{1/2} + 25.3851\ R_t + 14.0941\ R_t^{3/2} - 7.0261\ R_t^2 + 2.7081\ R_t^{5/2} + GS$$

$$R_t = C\ (S,t,0)/\ C\ (KCl,t,0)$$

GS = $[(t - 15)/ (1 + 0.0162(t - 15))] (0.0005 - 0.0056 R_t^{1/2} - 0.0066 R_t - 0.0375 R_t^{3/2} + 0.636 R_t^{2} - 0.0144 R_t^{5/2})$

2°CƒTƒ35°C

An extension of gives salinity at any pressure.

DEFINITION OF TEMPERATURE

Many physical processes depend on temperature; and a few can be used to define absolute temperature T. The unit of T is the kelvin, which has the symbol K. The fundamental processes used for defining an absolute temperature scale over the range of temperatures found in the ocean include: 1) the gas laws relating pressure to temperature of an ideal gas with corrections for the density of the gas; and 2) the voltage noise of a resistance R. The measurement of temperature using an absolute scale is difficult and the measurement is usually made by national standards laboratories. The absolute measurements are used to define a practical temperature scale based on the temperature of a few fixed points and interpolating devices which are calibrated at the fixed points.

For temperatures commonly found in the ocean, the interpolating device is a platinum-resistance thermometer. It consists of a loosely wound, strain-free, pure platinum wire whose resistance is a function of temperature. It is calibrated at fixed points between the triple point of equilibrium hydrogen at 13.8033K and the freezing point of silver at 961.78 K, including the triple point of water at 0.060°C, the melting point of Gallium at 29.7646°C, and the freezing point of Indium at 156.5985°C.

The triple point of water is the temperature at which ice, water, and water vapor are in equilibrium. The temperature scale in kelvin T is related to the temperature scale in degrees Celsius t [°C] by:

$$t\,[°C] = T\,[K] - 273.15$$

The practical temperature scale was revised in 1887, 1927, 1948, 1968, and 1990 as more accurate determinations of absolute temperature became accepted. The most recent scale is the International Temperature Scale of 1990. It differs slightly from the International Practical Temperature Scale of 1968 IPTS-68. At 0°C they are the same, and above 0°C ITS-90 is slightly cooler. $t_{90} - t_{68}$ = -0.002 at 10°C, -0.005 at 20°C, 0.007 at 30°C and -0.010 at 40°C. Notice that while oceanographers use thermometers calibrated with an accuracy of a millidegree, which is 0.001°C, the temperature scale itself has uncertainties of a few millidegrees.

GEOGRAPHICAL DISTRIBUTION OF SURFACE TEMPERATURE AND SALINITY

The distribution of temperature at the sea surface tends to be zonal, that is it tends to be independent of longitude. Warmest water is near the equator, coldest water is near the poles. The deviations from zonal are small.

Equatorward of 40°, cooler waters tend to be on the eastern side of the basin. North of this latitude, cooler waters tend to be on the western side. The anomalies of sea-surface temperature, the deviation from a long term average, are small, less than 1.5°C except in the equatorial Pacific where the deviations can be 3°C. The annual range of surface temperature is highest at mid-latitudes, especially on the western side of the ocean. In the west, cold air blows off the continents in winter and cools the ocean. The cooling dominates the heat budget. In the tropics the temperature range is mostly less than 2°C. Bottom: Annual range of sea-surface temperature in °C calculated from the Reynolds and Smith mean sea-surface temperature data set.

Contour interval is 1°C with heavy contours at 4°C and 8°C. Shaded areas exceed 8°C. The distribution of sea-surface salinity also tends to be zonal. The saltiest waters are at mid-latitudes where evaporation is high. Less salty waters are near the equator where rain freshens the surface water, and at high latitudes where melted sea ice freshens the surface waters. The zonal (east-west) average of salinity shows a close correlation between salinity and evaporation minus precipitation plus river input.

Because many large rivers drain into the Atlantic and the Arctic Sea, why is the Atlantic saltier than the Pacific? Broecker showed that 0.32 Sv of the water evaporated from the Atlantic does not fall as rain on land. Instead, it is carried by winds into the Pacific. Broecker points out that the quantity is small, equivalent to a little more than the flow in the Amazon River, but "were this flux not compensated by an exchange of more salty Atlantic waters for less salty Pacific waters, the salinity of the entire Atlantic would rise about 1 gram per liter per millennium."

Mean Temperature and Salinity of the Ocean The mean temperature of the ocean's waters is: t = 3.5°C; and the mean salinity is S = 34.7. The distribution about the mean is small: 50% of the water is in the range:

$$1.3°C < t < 3.8°C$$

$$34.6 < S < 34.8$$

THE OCEANIC MIXED LAYER AND THERMOCLINE

Wind blowing on the ocean stirs the upper layers leading to a thin mixed layer at the sea surface having constant temperature and salinity from the surface down to a depth where the values differ from those at the surface. The magnitude of the difference is arbitrary, but typically the temperature at the bottom of the layer must be no more than 0.02-0.1°C colder than at the surface.

Note that the both temperature and salinity must be constant in the mixed layer. The mixed layer is roughly 10-200 m thick over most of the tropical and mid-latitude belts. The depth and temperature of the mixed layer varies from day to day and from season to season in response to two processes:

1. Heat fluxes through the surface heat and cool the surface waters. Changes in temperature change the density contrast between the mixed layer and deeper waters. The greater the contrast, the more work is needed to mix the layer downward and visa versa.
2. Turbulence in the mixed layer mixes heat downward. The turbulence depends on the wind speed and on the intensity of breaking waves. Turbulence mixes water in the layer, and it mixes the water in the layer with water in the thermocline.

The mid-latitude mixed layer is thinnest in late summer when winds are weak, and sunlight warms the surface layer. At times, the heating is so strong, and the winds so weak, that the layer is only a few meters thick. In fall, the first storms of the season mix the heat down into the ocean thickening the mixed layer, but little heat is lost. In winter, heat is lost, and the mixed layer continues to thicken, becoming thickest in late winter. In spring, winds weaken, sunlight increases, and a new mixed layer forms.

Below the mixed layer, water temperature decreases rapidly with depth except at high latitudes. The range of depths where the rate of change, the gradient of temperature, is large is called the thermocline. Because density is closely related to temperature, the thermocline also tends to be the layer where density gradient is greatest, the pycnocline. The top of the thermocline varies slightly with the seasons. This is the seasonal thermocline. The permanent thermocline extends from below the seasonal thermocline to depths of 1500-2000m.

The mixed layer tends to be saltier than the thermocline between 10° and 40° latitude, where evaporation exceeds precipitation. At high latitudes the mixed layer is fresher because rain and melting ice reduce salinity. In some tropical regions, such as the warm pool in the western tropical Pacific, rain also produces a thin fresher mixed layer.

DENSITY, POTENTIAL TEMPERATURE, AND NEUTRAL DENSITY

During winter, cold water formed at the surface sinks to a depth determined by its density relative to the density of the deeper water. Currents then carry the water to other parts of the ocean. At all times, the water parcel moves to stay below less dense water and above more dense water. The distribution of currents within the ocean depends on the distribution of pressure, which depends on the variations of density inside the ocean as outlined in §10.4. So, if we want to follow water movement within the ocean, we need to know the distribution of density within the ocean.

Density and sigma-t

The calculation of water movement requires measurements of density with an accuracy of a few parts per million. This is not easy. Absolute Density of water can only be measured in special laboratories, and only with difficulty. The best accuracy is 1: $2.5 \times 10^5 = 4$ parts per million. To avoid the difficulty of working with absolute density, oceanographers use density relative to density of pure water. Density ñ is now defined using Standard Mean Ocean Water of known isotopic composition, assuming saturation of dissolved atmospheric gasses.

Here S, t, p refers to salinity, temperature, and pressure. In practice, density is not measured, it is calculated from in situ measurements of pressure, temperature, and conductivity using the equation of state for sea water. This can be done with an accuracy of two parts per million. Density of water at the sea surface is typically 1027 kg/m^3. For simplification, physical oceanographers often quote only the last 2 digits of the density, a quantity they call density anomaly or Sigma:

$$v\ (s,t,p) = u\ (s,t,p) - 1000\ kg/m^3$$

v (S, t, p) is typically 27.00 kg/m^3. The Working Group on Symbols in Oceanography recommends that v be replaced by o because v was originally defined relative to pure water and it was dimensionless. Here, however, we will follow common practice and use v. If we are studying surface layers of the ocean, we can ignore compressibility, and we use a new quantity sigma-t (written v_t):

$$v_t = v\ (S,t,0)$$

This is the density anomaly of a water sample when the total pressure on it has been reduced to atmospheric pressure (*i.e.* zero water pressure), but the temperature and salinity are in situ values.

Potential Temperature

As a water parcel moves within the ocean below the mixed layer, its salt and heat content can change only by mixing with other water. Thus we can use measurements of temperature and salinity to trace the path of the water. This is best done if we remove the effect of compressibility. As water sinks, pressure increases, the water is compressed, and the compression does work on the water. This increases the internal energy of the water. To understand how compression increases energy, consider a cube containing a fixed mass of water. As the cube sinks, its sides move inward as the cube is compressed. Recalling that work is force times distance, the work is the distance the side moves times the force exerted on the side by pressure. The change in internal energy may or may not result in a change in temperature. The internal energy of a fluid is the sum of molecular kinetic energy (temperature) and molecular potential energy.

To remove the influence of compressibility from measurements of temperature, oceanographers (and meteorologists who have the same problem in the atmosphere) use the concept of potential temperature. Potential temperature t is defined as the temperature of a parcel of water at the sea surface after it has been raised adiabatically from some depth in the ocean. Raising the parcel adiabatically means that it is raised in an insulated container so it does not exchange heat with its surroundings. Of course, the parcel is not actually brought to the surface. Potential temperature is calculated from the temperature in the water at depth, the in situ temperature.

Potential Density

If we are studying intermediate layers of the ocean, say at depths near a kilometer, we cannot ignore compressibility. Because changes in pressure primarily influence the temperature of the water, the influence of pressure can be removed, to a first approximation, by using the potential density. Potential density v_t is the density a parcel of water would have if it were raised adiabatically to the surface without change in salinity.

$$v_t = s(v, t, 0)$$

v_t is especially useful because it is a conserved thermodynamic property. Potential density is not useful for comparing density of water at great depths. If we bring water parcels to the surface and compare their densities, the calculation of potential density ignores the effect of pressure on the coeffcients for thermal and salt expansion.

As a result, two water samples having the same density but different temperature and salinity at a depth of four kilometers can have noticeably different potential density. In some regions the use of v (t) can lead to an apparent decrease of density with depth although we know that this is not possible because such a column of water would be unstable. To compare samples from great depths, it is better to bring both samples to a nearby depth instead of to the surface p = 0. For example, we can bring both parcels to a pressure of 4,000 decibars, which is near a depth of 4km:

$$v_4 = v(S, v, 4000)$$

where v_4 is the density of a parcel of water brought adiabatically to a pressure of 4,000 decibars. More generally, oceanographers sometimes use v_r

$$v_r = v(S, t, p, p_r)$$

where where p is pressure, and p_r is pressure at some reference level. In the level is p_r = 0 decibars, and in p_r = 4000 decibars.

The use of v_r leads to problems. If we wish to follow parcels of water deep in the ocean, we might use v_3 in some areas, and v_4 in others. But what happens when a parcel moves from a depth of 3 km in one area to a depth of 4 km in another?

There is a small discontinuity between the density of the parcel expressed as v_3 compared with density expressed as v_4. To avoid this difficulty, Jackett and McDougall (1997) proposed a new variable they called neutral density.

Neutral Surfaces and Density

A parcel of water moves locally along a path of constant density keeping below less dense water and above more dense water. More precisely, it moves along a path of constant potential density v_r referenced to the local depth r. Such a path is called a neutral path. A neutral surface element is the surface tangent to the neutral paths through a point in the water. No work is required to move a parcel on this surface because there is no buoyancy force acting on the parcel as it moves (if we ignore friction).

Now let's follow the parcel as it moves away from a local region. This can be done approximately but not exactly. At first we might think that because we know the tangents to the surface everywhere, we can define a surface that is the envelope of the tangents. But an exact surface is not mathematically possible in the real ocean, although we can come very close. Jackett and McDougall (1997) developed a practical neutral density variable $\tilde{a}^{n}$ and surface that stays within a few tens meters of an ideal surface anywhere in the world.

They constructed their variables using data in the Levitus (1982) atlas. The neutral density values were then used to label the data in the Levitus atlas. This prelabeled data set is used to calculate $\tilde{a}^n$ at new locations where t, S are measured as a function of depth by interpolation to the four closest points in the Levitus atlas.

Through this practice, neutral density $\tilde{a}^n$ is a function of salinity S, in situ temperature t, pressure p, longitude, and latitude. The neutral surface defined above differs only slightly from an ideal neutral surface. If a parcel moves around a gyre on the neutral surface and returns to its starting location, its depth at the end will differ by around 10m from the depth at the start. If potential density surfaces are used, the difference can be hundreds of meters, a far larger error.

Equation of state of sea water

Density of sea water is rarely measured. Density is calculated from measurements of temperature, conductivity, or salinity, and pressure using the equation of state of sea water.

The equation of state is an equation relating density to temperature, salinity, and pressure. The equation is derived by fitting curves through laboratory measurements of density as a function of temperature, pressure, and salinity, chlorinity, or conductivity.

The International Equation of State (1980) published by the Joint Panel on Oceanographic Tables and Standards (1981) is now used. The equation has an accuracy of 10 parts per million, which is 0.01 units of ó(è). I have not actually written out the equation of state because it consists of three polynomials with 41 constants.

Accuracy of Temperature, Salinity, and Density

If we want to distinguish between different water masses in the ocean, and if the total range of temperature and salinity is as small, then we must measure temperature, salinity, and pressure very carefully. We will need an accuracy of a few parts per million. Such accuracy can be achieved only if all quantities are carefully defined, if all measurements are made with great care, if all instruments are carefully calibrated, and if all work is done according to internationally accepted standards. The standards are laid out in Processing of Oceanographic Station Data published by UNESCO. The book contains internationally accepted definitions of primary variables such as temperature and salinity and methods for the measuring the primary variables. It also describes accepted methods for calculating quantities derived from primary variables, such as potential temperature, density, and stability.

MEASUREMENT OF TEMPERATURE

Temperature in the ocean has been measured many ways. Thermistors and mercury thermometers are commonly used on ships and buoys. These are calibrated in the laboratory before being used, and after use if possible, using mercury or platinum thermometers with accuracy traceable to national standards laboratories. Infrared radiometers on satellites measure the ocean's surface temperature.

Mercury Thermometer

This is the most widely used, non-electronic thermometer. It was widely used in buckets dropped over the side of a ship to measure the temperature of surface waters, on Nansen bottles to measure subsea temperatures, and in the laboratory to calibrate other thermometers. Accuracy is about ± 0.001°C with careful calibration.

One very important mercury thermometer is the reversing thermometer carried on Nansen bottles, which are described in the next section. It is a thermometer that has a constriction in the mercury capillary that causes the thread of mercury to break at a precisely determined point when the thermometer is turned upside down. The thermometer is lowered deep into the ocean in the normal position; and it is allowed to come to equilibrium with the water. Mercury expands into the capillary, and the amount of mercury in the capillary is proportional to temperature. The thermometer is then flipped upside down, the thread of mercury breaks trapping the mercury in the capillary, and the thermometer is brought back. The mercury in the capillary of the reversed thermometer is read on deck along with the temperature of a normal thermometer, which gives the temperature at which the reversed thermometer is read. The two readings give the temperature of the water at the depth where the thermometer was reversed.

The reversing thermometer is carried inside a glass tube which protects the thermometer from the ocean's pressure because high pressure can squeeze additional mercury into the capillary. If the thermometer is unprotected, the apparent temperature read on deck is proportional to temperature and pressure at the depth where the thermometer was flipped A pair of protected and unprotected thermometers gives temperature and pressure of the water at the depth the thermometer was reversed.

Pairs of reversing thermometers carried on Nansen bottles were the primary source of subsea measurements of temperature as a function of pressure from around 1900 to 1970.

Platinum Resistance Thermometer

This is the standard for temperature. It is used by national standards laboratories to interpolate between defined points on the practical temperature scale. It is used primarily to calibrate other temperature sensors.

Thermistor

A thermistor is a semiconductor having resistance that varies rapidly and predictably with temperature. It has been widely used on moored instruments and on instruments deployed from ships since about 1970. It has high resolution and an accuracy of about ± 0.001°C when carefully calibrated.

Bucket temperatures The temperature of surface waters has been routinely measured at sea by putting a mercury thermometer into a bucket which is lowered into the water, letting it sit at a depth of about a meter for a few minutes until the thermometer comes to equilibrium, then bringing it aboard and reading the temperature before water in the bucket has time to change temperature. The accuracy is around 0.1°C. This is a very common source of direct surface temperature measurements.

Ship Injection Temperature The temperature of the water drawn into the ship to cool the engines has been recorded routinely for decades. These recorded values of temperature are called injection temperatures. Errors are due to ship's structure warming water before it is recorded. This happens when the temperature recorder is not placed close to the point on the hull where water is brought in. Accuracy is 0.5°–1°C.

Advanced Very High Resolution Radiometer

The most commonly used instrument to measure sea-surface temperature from space is the Advanced Very High Resolution Radiometer AVHRR. The instrument has been carried on all polar-orbiting meteorological satellites operated by NOAA since Tiros-N was launched in 1978.

The instrument was originally designed to measure cloud temperatures and hence cloud height. The instrument had, however, sufficient accuracy and precision that it was soon used to measure regional and global temperature patterns at the sea surface. The instrument is a radiometer that converts infrared radiation into electrical signals. It includes a mirror that scans from side to side across the subsatellite track and reflects radiance from the ground into a telescope, a telescope that focuses the radiance on detectors, detectors sensitive to different wavelengths that convert the radiance at those wavelengths into electrical signals, and electronic circuitry to digitize and store the radiance values.

The instruments observes a 2700-km wide swath centered on the subsatellite track. Each observation along the scan is from a pixel that is roughly one kilometer in diameter near the center of the scan and that increases in size with distance from the subsatellite track. The radiometers measures infrared radiation emitted from the surface in five wavelength bands: three infrared bands: 3.55–3.93 μm, 10.30–11.30 μm, and 11.50–12.50 μm; a near-infrared band at 0.725–1.00 μm; and a visible-light band at 0.58–0.68 μm. All infrared bands include radiation emitted from the sea and from water vapor in the air along the path from the satellite to the ground. The 3.7 μm band is least sensitive to water vapor and other errors, but it works only at night because sunlight has radiance in this band. The two longest wavelength bands at 10.8 μm and 12.0 μm are used to observe sea-surface temperature and water vapor along the path in daylight.

Data with 1-km resolution are transmitted directly to ground stations that view the satellite as it passes the station. This is the Local Area Coverage mode. Data are also averaged to produce observations from 4km × 4km pixels. These data are stored on tape recorders and later transmitted to NOAA receiving stations. This is the Global Area Coverage mode.

The swath width is sufficiently wide that the satellite views the entire Earth twice per day, at approximately 09:00 AM and 9:00 PM local time. Areas at high latitudes may be observed as often as eight or more times per day. The most important errors are due to:

1. Unresolved or undetected clouds: Large, thick clouds are obvious in the images of water temperature Thin clouds such as low stratus and high cirrus produce much small errors that are difficult or almost impossible to detect. Clouds smaller in diameter than 1km, such as trade-wind cumuli, are also difficult to detect. Special techniques have been developed for detecting small clouds.
2. Water vapor, which absorbs part of the energy radiated from the sea surface: Water vapor reduces the apparent temperature of the sea surface. The influence is different in the 10.8 μm and 12.0 μm channels, allowing the difference in the two signals to be used to reduce the error.
3. Aerosols, which absorb infrared radiation. They radiate at temperatures found high in the atmosphere. Stratospheric aerosols generated by volcanic eruptions can lower the observed

temperatures by up to a few degrees Celsius. Dust particles carried over the Atlantic from Saharan dust storms can also cause errors.

4. Skin temperature errors. The infrared radiation seen by the instrument comes from a layer at the sea surface that is only a few micrometers thick. The temperature in this layer is not quite the same as temperature a meter below the sea surface. They can differ by several degrees when winds are light. This error is greatly reduced when AVHRR data are used to interpolate between ship measurements of surface temperature.

Maps of temperature processed from Local Area Coverage of cloud-free regions show variations of temperature with a precision of 0.1°C. These maps are useful for observing local phenomena including patterns produced by local currents.

Global maps are made by the U. S. Naval Oceanographic Office, which receives the global AVHRR data directly from NOAA's National Environmental Satellite, Data and Information Service in near-real time each day. The data are carefully processed to remove the influence of clouds, water vapor, aerosols, and other sources of error. Data are then used to produce global maps between ± 70° with an accuracy of ± 0.6°C. The maps of sea-surface temperature are sent to the U.S. Navy and to NOAA's National Centers for Environmental Prediction. In addition, the office produces daily 100km global and 14km regional maps of temperature.

Global Maps of Sea-Surface Temperature

Global, monthly maps of surface temperature are produced by the National Centers for Environmental Prediction using Reynolds optimal-interpolation method. The technique blends ship and buoy measurements of sea-surface temperature with AVHRR data processed by the Naval Oceanographic Office in 1° areas for a month. Essentially, AVHRR data are interpolated between buoy and ship reports using previous information about the temperature field. Overall accuracy ranges from approximately ± 0.3°C in the tropics to ± 0.5°C near western boundary currents in the northern hemisphere where temperature gradients are large. Maps are available from November 1981.

Maps of mean temperature have also been made from ICOADS data. The data are poorly distributed in time and space except for some areas of the northern hemisphere. In addition, Reynolds and Smith found that ship temperature data had errors twice as large as temperature errors in data from buoys and AVHRR. Thus, space data processed by Reynolds are more accurate, and better distributed than ICOADS.

Anomalies of sea-surface temperature are calculated using mean sea-surface temperature from the period 1950-979 calculated from ICOADS supplemented with four years of satellite data 1982-1985.

MEASUREMENT OF CONDUCTIVITY

Conductivity is measured by placing platinum electrodes in seawater and measuring the current that flows when there is a known voltage between the electrodes. The current depends on conductivity, voltage, and volume of sea water in the path between electrodes. If the elecrodes are in a tube of non-conducting glass, the volume of water is accurately known, and the current is independent of other objects near the conductivity cell. The best measurements of salinity from conductivity give salinity with an accuracy of ±0.005. Before conductivity measurements were widely used, salinity was measured using chemical titration of the water sample with silver salts. The best measurements of salinity from titration give salinity with an accuracy of ±0.02.

Individual salinity measurements are calibrated using standard seawater. Long-term studies of accuracy use data from measurements of deep water masses of known, stable, salinity. For example, Saunders (1986) noted that temperature is very accurately related to salinity for a large volume of water contained in the deep basin of the northwest Atlantic under the Mediterranean out flow. He used the consistency of measurements of temperature and salinity made at many hydrographic stations in the area to estimate the accuracy of temperature, salinity and oxygen measurements. He concluded that the most careful measurements made since 1970 have an accuracy of 0.005 for salinity and 0.005°C for temperature. The largest source of salinity error was the error in determination of the standard water used for calibrating the salinity measurements.

Gouretski and Jancke (1995) estimated accuracy of salinity measurements as a function of time. Using high quality measurements from 16,000 hydrographic stations in the south Atlantic from 1912 to 1991, they estimated accuracy by plotting salinity as a function of temperature using all data collected below 1500m in twelve regions for each decade from 1920 to 1990. A plot of accuracy as a function of time since 1920 shows consistent improvement in accuracy since 1950. Recent measurements of salinity are the most accurate.

The standard deviation of modern salinity data collected from all areas in the South Atlantic from 1970 to 1993 adjusted as described by Gouretski and Jancke (1995) was 0.0033. More recent instruments such as the Sea-Bird Electronics Model 911 Plus have an accuracy of better than 0.005psu without adjustments. A careful comparison of salinity measured at 43° 10′N, 14° 4.5′W by the 911 Plus with historic data collected by Saunders (1986) gives an accuracy of 0.002.

MEASUREMENT OF PRESSURE

Pressure is routinely measured by many different types of instruments.

The SI unit of pressure is the pascal (Pa), but oceanographers normally report pressure in decibars (dbar), where:

1 dbar = 10^4 Pa

because the pressure in decibars is almost exactly equal to the depth in meters. Thus 1000 dbar is the pressure at a depth of about 1000 m.

- Strain Gage: This is the simplest and cheapest instrument, and it is widely used. Accuracy is about ±1%.
- Vibratron: Much more accurate measurements of pressure can be made by measuring the natural frequency of a vibrating tungsten wire stretched in a magnetic field between diaphragms closing the ends of a cylinder. Pressure distorts the diaphragm, which changes the tension on the wire and its frequency. The frequency can be measured from the changing voltage induced as the wire vibrates in the magnetic field. Accuracy is about ±0.1%, or better when temperature controlled. Precision is 100–1000 times better than accuracy. The instrument is used to detect small changes in pressure at great depths. Snodgrass (1968) obtained a precision equivalent to a change in depth of ±0.8 mm at a depth of 3 km.
- Quartz crystal: Very accurate measurements of pressure can also be made by measuring the natural frequency of a quartz crystal cut for minimum temperature dependence. The best accuracy is obtained when the temperature of the crystal is held constant. The accuracy is ±0.015%; and precision is ±0.001% of full-scale values.
- Quartz Bourdon Gage: Accuracy and stability comparable to quartz crystals. It too is used for long-term measurements of pressure in the deep sea.

MEASUREMENT OF TEMPERATURE AND SALINITY WITH DEPTH

Temperature, salinity, and pressure are measured as a function of depth using various instruments or techniques, and density is calculated from the measurements.

Bathythermograph (BT)

A mechanical device that measured temperature vs depth on a smoked glass slide. The device was widely used to map the thermal structure of the upper ocean, including the depth of the mixed layer before being replaced by the expendable bathythermograph in the 1970s.

Expendable Bathythermograph (XBT)

An electronic device that measures temperature vs depth using a thermistor on a free-falling streamlined weight. The thermistor is connected to an ohm-meter on the ship by a thin copper wire that is spooled out from the sinking weight and from the moving ship. The XBT is now the most widely used instrument for measuring the thermal structure of the upper ocean. Approximately 65,000 are used each year. The streamlined weight falls through the water at a constant velocity. So depth can be calculated from fall time with an accuracy of ±2%. Temperature accuracy is ±0.1°C. And, vertical resolution is typically 65 cm. Probes reach to depths of 200 m to 1830 m depending on model.

Nansen Bottles

Deployed from ships stopped at hydrographic stations. Hydrographic stations are places where oceanographers measure water properties from the surface to some depth, or to the bottom, using instruments lowered from a ship. Usually 20 bottles were attached at intervals of a few tens to hundreds of meters to a wire lowered over the side of the ship. The distribution with depth was selected so that most bottles are in the upper layers of the water column where the rate of change of temperature in the vertical is greatest. A protected reversing thermometer for measuring temperature was attached to each bottle along with an unprotected reversing thermometer for measuring depth. The bottle contains a tube with valves on each end to collect sea water at depth. Salinity was determined by laboratory analysis of water sample collected at depth.

After bottles had been attached to the wire and all had been lowered to their selected depths, a lead weight was dropped down the wire. The weight tripped a mechanism on each bottle, and the bottle flipped over, reversing the thermometers, shutting the valves and trapping water in the tube, and releasing another weight. When all bottles had been tripped, the string of bottles was recovered. The deployment and retrieval typically took several hours.

CTD

Mechanical instruments on Nansen bottles were replaced beginning in the 1960s by an electronic instrument, called a CTD, that measured conductivity, temperature, and depth. The measurements are recorded in digital form either within the instrument as it is lowered from a ship or on the ship. Temperature is usually measured by a thermistor; conductivity is measured by induction; pressure is measured by a quartz crystal.

LIGHT IN THE OCEAN AND ABSORPTION OF LIGHT

Sunlight in the ocean is important for many reasons: It heats sea water, warming the surface layers; it provides energy required by phytoplankton; it is used for navigation by animals near the surface; and reflected subsurface light is used for mapping chlorophyll concentration from space. Light in the ocean travels at a velocity equal to the velocity of light in a vacuum divided by the index of refraction (n), which is typically n = 1.33. Hence the velocity in water is about 2.25×108 m/s. Because light travels slower in water than in air, some light is reflected at the sea surface. For light shining straight down on the sea, the reflectivity is (n - 1)2/(n + 1)2. For seawater, the reflectivity is 0.02 = 2%. Hence most sunlight reaching the sea surface is transmitted into the sea, little is reflected.

This means that sunlight incident on the ocean in the tropics is mostly absorbed below the sea surface. The rate at which sunlight is attenuated determines the depth which is lighted and heated by the sun. Attenuation is due to absorption by pigments and scattering by molecules and particles. Attenuation depends on wavelength. Blue light is absorbed least, red light is absorbed most strongly. Attenuation per unit distance is proportional to the radiance or the irradiance of light:

where x is the distance along beam, c is an attenuation coefficient and I is irradiance or irradiance. Radiance is the power per unit area per solid angle. It is useful for describing the energy in a beam of light coming from a particular direction. Sometimes we want to know how much light reaches some depth in the ocean regardless of which direction it is going. In this case we use irradiance, which is the power per unit area of surface. If the absorption coefficient is constant, the light intensity decreases exponentially with distance.

$$I_2 = I_1 \exp(-cx)$$

where I_1 is the original radiance or irradiance of light, and I_2 is the radiance or irradiance of light after absorption.

Clarity of Ocean Water Sea water in the middle of the ocean is very clear—clearer than distilled water. These waters are a very deep, cobalt, blue—almost black. Thus the strong current which flows northward offshore of Japan carrying very clear water from the central Pacific into higher latitudes is known as the Black Current, or Kuroshio in Japanese. The clearest ocean water is called Type I waters by Jerlov. The water is so clear that 10% of the light transmitted below the sea surface reaches a depth of 90m.

In the subtropics and mid-latitudes closer to the coast, sea water contains more phytoplankton than the very clear central-ocean waters. Chlorophyll pigments in phytoplankton absorb light, and the plants themselves scatter light. Together, the processes change the color of the ocean as seen by observer looking downward into the sea. Very productive waters, those with high concentrations of phytoplankton, appear blue-green or green. On clear days the color can be observed from space. This allows ocean-color scanners, such as those on SeaWiFS, to map the distribution of phytoplankton over large areas.

As the concentration of phytoplankton increases, the depth where sunlight is absorbed in the ocean decreases. The more turbid tropical and mid-latitude waters are classi.ed as type II and III waters by Jerlov. Thus the depth where sunlight warms the water depends on the productivity of the waters. This complicates the calculation of solar heating of the mixed layer. Coastal waters are much less clear than waters offshore. Measurement of Chlorophyll from Space: The color of the ocean, and hence the chlorophyll concentration in the upper layers of the ocean has been measured by the Coastal Zone Color Scanner carried on the Nimbus-7 satellite launched in 1978 and by the Sea-viewing Wide Field-of-view Sensor carried on SeaStar, launched in 1997. The latter instrument measures upwelling radiance in eight wavelength bands between 412nm and 856nm.

Most of the upwelling radiance seen by the satellite comes from the atmosphere. Only about 10% comes from the sea surface. Both air molecules and aerosols scatter light; and very accurate techniques have been developed to remove the influence of the atmosphere.

The total radiance L_t received by an instrument in space is:

$$L_t\ (o_i) = t\ (o_i)\ L_W\ (o_i) + L_r\ (o_i) + L_a\ (o_i)$$

where o_i is the wavelength of the radiation in the band measured by the instrument, L_W is the radiance leaving the sea surface, L_r is radiance scattered by molecules, called the Rayleigh radiance, L_a is radiance scattered from aerosols, and t is the transmittance of the atmosphere. L_r can be calculated from theory; and L_a can be calculated from the amount of red light received at the instrument because very little red light is reflected from the water. Therefore L_W can be calculated from the radiance measured at the spacecraft.

Chlorophyll concentration in the water column is calculated from the ratio of L_W at two frequencies. Using data from the Coastal Zone Color Scanner, Gordon et al., (1983) proposed where C is the chlorophyll concentration in the surface layers in mg pigment/m^3, and $L_W(443)$, $L_W(520)$, and $L_W(550)$ is the radiance at wavelengths of 443 nm, 520nm, and 550nm. C_{13} is used when C_{13} *f* 1.5 mg/m^3; otherwise C_{23} is used. The technique is used to calculate chlorophyll concentration within a factor of 50% over a wide range of concentrations from 0.01 mg/m^3 to 10 mg/m^3.

10

Climate Change Impacts on Coastal and Marine Engineering

TEMPERATURE AND PRESSURE BELTS OF THE WORLD

TEMPERATURE

Temperature is a physical property of matter that quantitatively expresses the common notions of hot and cold. Objects of low temperature are cold, while various degrees of higher temperatures are referred to as warm or hot. Quantitatively, temperature is measured with thermometers, which may be calibrated to a variety of temperature scales.

Much of the world uses the Celsius scale for most temperature measurements. It has the same incremental scaling as the Kelvin scale used by scientists, but fixes its null point, at 0°C = 273.15K, the freezing point of water. A few countries, most notably the United States, use the Fahrenheit scale for common purposes, a historical scale on which water freezes at 32 °F and boils at 212 °F.

For practical purposes of scientific temperature measurement, the International System of Units defines a scale and unit for the thermodynamic temperature by using the easily reproducible temperature of the triple point of water as a second reference point. For historical reasons, the triple point is fixed at 273.16 units of the measurement increment, which has been named the kelvin in honor of the Scottish physicist who first defined the scale. The unit symbol of the kelvin is K.

Temperature is one of the principal properties studied in the field of thermodynamics. Particularly important in this field are the differences in temperature between regions of matter, because such differences are the driving force for heat, which is the transfer of thermal energy. Spontaneously, heat flows only from regions of higher temperature to regions of lower temperature. If no heat is transferred between two objects, the objects have the same temperature.

In the classical thermodynamic approach to temperature, temperature of an object varies with the speed of the particles it contains, raised to the second power. Therefore, temperature is tied directly to the mean kinetic energy of particles moving relative to the center of mass coordinates for that object.

Temperature is an intensive variable because it is independent of the bulk amount of elementary entities contained inside, be they atoms, molecules, or electrons. In order for the temperature of a system to be defined, the system must be in thermodynamic equilibrium. Temperature may be considered to vary with position only if, for every point, there is a small neighborhood around that point can be treated as a thermodynamic system in equilibrium. In the statistical thermodynamic approach, degrees of freedom are used instead of particles.

In a more fundamental approach, the empirical definition of temperature arises from the conditions of thermal equilibrium, expressed as the zeroth law of thermodynamics. When two systems are in thermal equilibrium, they have the same temperature, which is also a matter of common experience. The extension of this principle as an equivalence relation between multiple systems fundamentally justifies the use of a thermometer and prescribes the principles of its construction to measure temperature. While the zeroth law permits the definition of a set of many empirical scales of temperature, the second law of thermodynamics selects the definition of a single preferred, absolute temperature function, whence called the thermodynamic temperature.

This function is the variation of the internal energy with respect to changes in the entropy of a system. Its natural, intrinsic origin or null point is absolute zero at which the entropy of any system is at a minimum. Although this is the lowest absolute temperature described by the model, the third law of thermodynamics postulates that absolute zero cannot be attained by any physical system. Macroscopically, temperature is related to the thermal energy held by matter. An immediate way of sensing this is by touching the material and deciding whether it is hot, warm, or cold.

A thermometer precisely measures the temperature and indicates a numerical value for the temperature. On the molecular level, temperature is the result of the motion of the particles that constitute the material. Moving particles carry kinetic energy. Temperature increases as this motion and the kinetic energy increase. The motion may be the translational motion of particles, or the energy of the particle due to molecular vibration or the excitation of an electron energy level.

Although very specialized laboratory equipment is required to directly detect the translational thermal motions, thermal collisions by atoms or molecules with small particles suspended in a fluid produces Brownian motion that can be seen with an ordinary microscope. The thermal motions of atoms are very fast and temperatures close to absolute zero are required to directly observe them. For instance, when scientists at the NIST achieved a record-setting low temperature of 700 nK in 1994, they used laser equipment to create an optical lattice to adiabatically cool caesium atoms. They then turned off the entrapment lasers and directly measured atom velocities of 7mm per second in order to calculate their temperature.

Molecules, such as oxygen have more degrees of freedom than single spherical atoms: they undergo rotational and vibrational motions as well as translations. Heating results in an increase in temperature due to an increase in the average translational energy of the molecules. Heating will also cause, through equipartitioning, the energy associated with vibrational and rotational modes to increase. Thus a diatomic gas will require a higher energy input to increase its temperature by a certain amount, *i.e.* it will have a higher heat capacity than a monatomic gas.

The process of cooling involves removing thermal energy from a system. When no more energy can be removed, the system is at absolute zero, which cannot be achieved experimentally. Absolute zero is the null point of the thermodynamic temperature scale, also called absolute temperature. If it were possible to cool a system to absolute zero, all motion of the particles comprising matter would cease and they would be at complete rest in this *classical* sense. Microscopically in the description of quantum mechanics, however, matter still has zero-point energy even at absolute zero, because of the uncertainty principle. Absolute zero is defined as a temperature of precisely 0 kelvins, which is equal to –273.15 °C or –459.68 °F.

Use in science

Temperature plays an important role in all fields of natural science, including physics, geology, chemistry, atmospheric sciences and biology. Many physical properties of materials including the phase, density, solubility, vapor pressure, and electrical conductivity depend on the temperature. Temperature also plays an important role in determining the rate and extent to which chemical reactions occur.

This is one reason why the human body has several elaborate mechanisms for maintaining the temperature at 310 K, since temperatures only a few degrees higher can result in harmful reactions with serious consequences. Temperature also controls the thermal radiation emitted from a surface. One application of this effect is the incandescent light bulb, in which a tungsten filament is electrically heated to a temperature at which significant quantities of visible light are emitted. For study and analysis, real world systems are usually spatially and temporally divided into cells of small size, in which classical thermodynamical equilibrium conditions for matter are fulfilled to good approximation.

Foundation

Temperature may be viewed as a quality of heat, as distinct from a quantity of heat· Mathematically, this quality exists on a one-dimensional manifold. Physically, this means that, when comparing the temperatures of any two given systems in their respective separate thermodynamic equilibria, any two proper thermometers will agree as which is the hotter of the two given systems, or that they have the same temperature. This does not require the two thermometers to have a linear relation between their numerical scale readings, but it does require that the relation between their numerical readings shall be strictly monotonic. This is a fundamental character of temperature and thermometers.

Macroscopic and statistical thermodynamics give a formal account of temperature. Temperature is often discussed in modern theoretical texts in terms of the zeroth law of thermodynamics; long before the statement of the "zeroth law", experimental physicists, for example Galileo and Newton, found that there are indefinitely many empirical temperature scales. Amongst all possible temperature scales, the second law distinguishes absolute temperature scales. Statistical thermodynamics provides a detailed microscopic explanation based on the probabilistic properties of large ensembles of particles.

Temperature is a property determined by the average of a property of many particles and is defined in a state of at least local thermodynamic equilibrium, unlike other thermodynamic quantities, such as entropy and heat, whose microscopic definitions are valid even far away from equilibrium.

Except for a system undergoing a first-order phase change such as the melting of ice, as a system receives heat, its temperature will rise. For a system undergoing such a phase change, its temperature will remain constant as the latent heat in the system increases. Conversely, a loss of heat from a system will usually decrease its temperature. When two systems are at the same temperature, no heat transfer can occur between them. When a temperature difference does exist, heat will transfer from the warmer system to the colder system, until they are at thermal equilibrium. This transfer occurs via heat conduction.

Statistically, temperature (T) is a direct measure of the mean kinetic energy of the particles forming a sample of matter. For each degree of freedom that a particle possesses, the mean kinetic energy (E_k) of the particles is:

$$\bar{E}_k \ @ \ \frac{1}{2}kT,$$

where k is the Boltzmann constant, a fixed proportionality factor introduced by the system of units used to measure energy and temperature. Macroscopically, temperature is related to the amount of internal energy and enthalpy of a system: the higher the temperature of a system, the higher its internal energy and enthalpy. For a system in thermal equilibrium at a constant volume, temperature is thermodynamically defined in terms of its energy (E) and entropy (S) as:

$$T = \left(\frac{\partial S}{\partial E}\right)^{-1}$$

Temperature is an intensive property of a system, meaning that it does not depend on the system size, the amount or type of material in the system, the same as for the pressure and density. By contrast, mass, volume, and entropy are extensive properties, and depend on the amount of material in the system.

Heat capacity

When a sample is heated, meaning it receives thermal energy from an external source, some of the introduced heat is converted into kinetic energy, the rest to other forms of internal energy, specific to the material. The amount converted into kinetic energy causes the temperature of the material to rise. The introduced heat (δQ) divided by the observed temperature change is the heat capacity (C) of the material.

$$C = \frac{\delta Q}{\delta T}$$

If heat capacity is measured for a well defined amount of substance, the specific heat is the measure of the heat required to increase the temperature of such a unit quantity by one unit of temperature. For example, to raise the temperature of water by one kelvin requires 4186 joules per kilogram.

Temperature measurement

Temperature measurement using modern scientific thermometers and temperature scales goes back at least as far as the early 18th century, when Gabriel Fahrenheit adapted a thermometer and a scale both developed by Ole Christensen Rømer. Fahrenheit's scale is still in use in the United States for non-scientific applications.

Temperature is measured with thermometers that may be calibrated to a variety of temperature scales. In most of the world, the Celsius scale is used for most temperature measuring purposes. Most scientist measures temperature using the Celsius scale and the thermodynamic temperature using the Kelvin scale, which is the Celsius scale offset so that its null point is 0K = −273.15°C, or absolute zero. Many engineering fields in the U.S., notably high-tech and US federal specifications also use the Kelvin and Celsius scales. Other engineering fields in the U.S. also rely upon the Rankine scale when working in thermodynamic-related disciplines such as combustion.

Units

The basic unit of temperature in the International System of Units is the kelvin. It has the symbol K. For everyday applications, it is often convenient to use the Celsius scale, in which 0°C corresponds very closely to the freezing point of water and 100°C is its boiling point at sea level. Because liquid droplets commonly exist in clouds at sub-zero temperatures, 0°C is better defined as the melting point of ice. In this scale a temperature difference of 1 degree Celsius is the same as a 1kelvin increment, but the scale is offset by the temperature at which ice melts.

By international agreement the Kelvin and Celsius scales are defined by two fixing points: absolute zero and the triple point of Vienna Standard Mean Ocean Water, which is water specially prepared with a specified blend of hydrogen and oxygen isotopes. Absolute zero is defined as precisely 0K and "273.15°C. It is the temperature at which all classical translational motion of the particles comprising matter ceases and they are at complete rest in the classical model. Quantum-mechanically, however, zero-point motion remains and has an associated energy, the zero-point energy. Matter is in its ground state, and contains no thermal energy.

The triple point of water is defined as 273.16K and 0.01°C. This definition serves the following purposes: it fixes the magnitude of the kelvin as being precisely 1 part in 273.16 parts of the difference between absolute zero and the triple point of water; it establishes that one kelvin has precisely the same magnitude as one degree on the Celsius scale; and it establishes the difference between the null points of these scales as being 273.15K. In the United States, the Fahrenheit scale is widely used. On this scale the freezing point of water corresponds to 32 °F and the boiling point to 212 °F. The Rankine scale, still used in fields of chemical engineering in the U.S., is an absolute scale based on the Fahrenheit increment.

Plasma physics

The field of plasma physics deals with phenomena of electromagnetic nature that involve very high temperatures. It is customary to express temperature in electronvolts (eV) or kiloelectronvolts (keV), where 1 eV = 11605K. In the study of QCD matter one routinely encounters temperatures of the order of a few hundred MeV, equivalent to about 10^{12}K.

Theoretical foundation

Historically, there are several scientific approaches to the explanation of temperature: the classical thermodynamic description based on empirical variables that can be measured in a laboratory, the kinetic theory of gases which relates the macroscopic description to the probability distribution of the energy of motion of gas particles, and a microscopic explanation based on statistical physics and quantum theory. In addition, rigorous and purely mathematical treatments have provided an axiomatic approach to classical thermodynamics and temperature. Statistical physics provides a deeper understanding by describing the atomic behavior of matter, and derives macroscopic properties from statistical averages of microscopic states, including both classical and quantum states. In the fundamental physical description, using natural units, temperature may be measured directly in units of energy. However, in the practical systems of measurement for science, technology, and commerce, such as the modern metric system of units, the macroscopic and the microscopic descriptions are interrelated by the Boltzmann constant, a proportionality factor that scales temperature to the microscopic mean kinetic energy.

The microscopic description in statistical mechanics is based on a model that decomposes a system into its fundamental particles of matter or into a set of classical or quantum-mechanical oscillators and considers the system as a statistical ensemble of microstates. As a collection of classical material particles, temperature is a measure of the mean energy of motion, called kinetic energy, of the particles, whether in solids, liquids, gases, or plasmas. Kinetic energy, a concept of classical mechanics, is one half the product of mass and the square of a particle's velocity. In this mechanical interpretation of thermal motion, the kinetic energies of material particles may reside in the velocity of the particles of their translational or vibrational motion or in the inertia of their rotational modes. In monoatomic perfect gases and, approximately, in most gases, temperature is a measure of the mean particle kinetic energy. It also determines the probability distribution function of the energy. In condensed matter, and particularly in solids, this purely mechanical description is often less useful and the oscillator model provides a better description to account for quantum mechanical phenomena. Temperature determines the statistical occupation of the microstates of the ensemble. The microscopic definition of temperature is only meaningful in the thermodynamic limit, meaning for large ensembles of states or particles, to fulfill the requirements of the statistical model.

In the context of thermodynamics, the kinetic energy is also referred to as thermal energy. The thermal energy may be partitioned into independent components attributed to the degrees of freedom of the particles or to the modes of oscillators in a thermodynamic system. In general, the number of these degrees of freedom that are available for the equipartitioning of energy depend on the temperature, *i.e.* the energy region of the interactions under consideration. For solids, the thermal energy is associated primarily with the vibrations of its atoms or molecules about their equilibrium position. In an ideal monatomic gas, the kinetic energy is found exclusively in the purely translational motions of the particles. In other systems, vibrational and rotational motions also contribute degrees of freedom.

Kinetic theory of gases

The kinetic theory of gases uses the model of the ideal gas to relate temperature to the average kinetic energy of the atoms in a container of gas. Classical mechanics defines the kinetic energy as follows:

$$E_k @ \frac{1}{2} m v^2 ,$$

Where m is the particle mass and v its velocity. The distribution of energies of the particles in any gas are given by the Maxwell-Boltzmann distribution. The temperature of a classical ideal gas is related to its average kinetic energy via the equation:

$$\bar{E}_k @ \frac{1}{2} kT ,$$

for each degree of freedom, where $k = R/\ n$ (n= Avogadro number, R= ideal gas constant). This relation is valid in the classical regime, *i.e.* when the particle density is much less than $1/\ o^3$, where o is the thermal de Broglie wavelength. A monoatomic gas has only the three translational degrees of freedom.

The second law of thermodynamics states that any two given systems when interacting with each other will later reach the same average energy per particle and hence the same temperature. In a mixture of particles of various masses, the heaviest particles will move slower than lighter particles, but have the same average kinetic energy. A neon atom moves slower relative to a hydrogen molecule of the same kinetic energy; a pollen particle suspended in water moves in a slow Brownian motion among fast moving water molecules.

Zeroth law of thermodynamics

If two systems of different temperatures are brought into thermal contact, they exchange heat accompanied by changes of other state variables. The system achieve a state of thermal equilibrium in which no further changes occur. The mathematical description of thermal equilibrium forms a basis for the definition of temperature. This has been formulated as the zeroth law of thermodynamics which states that if two systems are each in thermal equilibrium with a third system, then they are also in thermal equilibrium with each other.

Since all three are in thermal equilibrium, it is reasonable to postulate an empirical property that has the same value for all systems. Temperature is just such a property. The transitive relation of thermal equilibrium permits the construction of a thermometer to measure this property. For suitable systems, an empirical temperature scale may be defined by the variation of one of the other state variables, such as pressure, when all other coordinates are fixed. The second law of thermodynamics is used to define an absolute thermodynamic temperature scale, which is what is commonly referred to as "temperature". A temperature scale is based on the properties of some reference system to which other thermometers may be calibrated. One such reference system is a fixed quantity of gas. The ideal gas law indicates that the product of the pressure (p) and volume (V) of a gas is directly proportional to the thermodynamic temperature:

$$pV @ nRT$$

where T is temperature, n is the number of moles of gas and R = 8.314472(15) $\text{Jmol}^{-1}\text{K}^{-1}$ is the gas constant. Reformulating the pressure-volume term as the sum of classical mechanical particle energies in terms of particle mass, m, and root-mean-square particle speed v, the ideal gas law directly provides the relationship between kinetic energy and temperature:

$$\frac{1}{2}mv_{rms}^2 @ \frac{3}{2}kT.$$

Thus, one can define a scale for temperature based on the corresponding pressure and volume of the gas: the temperature in kelvins is the pressure in pascals of one mole of gas in a container of one cubic metre, divided by the gas constant. In practice, such a gas thermometer is not very convenient, but other thermometers can be calibrated to this scale.

The pressure, volume, and the number of moles of a substance are all inherently greater than or equal to zero, suggesting that temperature must also be greater than or equal to zero. As a practical matter it is not possible to use a gas thermometer to measure absolute zero temperature since the gasses tend to condense into a liquid long before the temperature reaches zero. It is possible, however, to extrapolate to absolute zero by using the ideal gas law.

Second law of thermodynamics

In the previous section temperature was defined in terms of the zeroth law of thermodynamics. It is also possible to define temperature in terms of the second law of thermodynamics which deals with entropy. Entropy is a measure of the disorder in a system. The second law states that any process will result in either no change or a net increase in the entropy of the universe. This can be understood in terms of probability.

For example, in a series of coin tosses, a perfectly ordered system would be one in which either every toss comes up heads or every toss comes up tails. This means that for a perfectly ordered set of coin tosses, there is only one set of toss outcomes possible: the set in which 100% of tosses come up the same. On the other hand, there are multiple combinations that can result in disordered or mixed systems, where some fraction are heads and the rest tails. A disordered system can be 90% heads and 10% tails, or it could be 98% heads and 2% tails, et cetera. As the number of coin tosses increases, the number of possible combinations corresponding to imperfectly ordered systems increases. For a very large number of coin tosses, the combinations to ~50% heads and ~50% tails dominates and obtaining an outcome significantly different from 50/50 becomes extremely unlikely. Thus the system naturally progresses to a state of maximum disorder or entropy.

It has been previously stated that temperature controls the flow of heat between two systems and was just shown that the universe tends to progress so as to maximize entropy, which is expected of any natural system. Thus, it is expected that there is some relationship between temperature and entropy. To find this relationship, the relationship between heat, work and temperature is first considered.

A heat engine is a device for converting thermal energy into mechanical energy, resulting in the performance of work, and analysis of the Carnot heat engine provides the necessary relationships. The work from a heat engine corresponds to the difference between the heat put into the system at the high temperature, q_H and the heat ejected at the low temperature, q_C. The efficiency is the work divided by the heat put into the system or:

$$\text{efficiency} @ \frac{w_{cy}}{q_H} @ \frac{q_H \; 0 \; q_C}{q_H} @ 1 \; 0 \; \frac{q_C}{q_H}$$

where w_{cy} is the work done per cycle. The efficiency depends only on q_C/q_H. Because q_C and q_H correspond to heat transfer at the temperatures T_C and T_H, respectively, q_C/q_H should be some function of these temperatures:

$$\frac{q_C}{q_H} @ f(T_H, T_C)$$

Carnot's theorem states that all reversible engines operating between the same heat reservoirs are equally efficient. Thus, a heat engine operating between T_1 and T_3 must have the same efficiency as one consisting of two cycles, one between T_1 and T_2, and the second between T_2 and T_3. This can only be the case if:

$$q_{13} @ \frac{q_1 q_2}{q_2 q_3}$$

which implies:

$$q_{13} = f(T_1, T_3) = f(T_1, T_2) f(T_2, T_3)$$

Since the first function is independent of T_2, this temperature must cancel on the right side, meaning $f(T_1,T_3)$ is of the form $g(T_1)/g(T_3)$ (*i.e.* $f(T_1,T_3) = f(T_1,T_2)f(T_2,T_3) = g(T_1)/g(T_2) \cdot g(T_2)/g(T_3) = g(T_1)/g(T_3)$), where g is a function of a single temperature. A temperature scale can now be chosen with the property that:

$$\frac{q_C}{q_H} @ \frac{T_C}{T_H}$$

Substituting Equation 4 back into Equation 2 gives a relationship for the efficiency in terms of temperature:

$$\text{efficiency} @ 1\,0\,\frac{q_C}{q_H} @ 1\,0\,\frac{T_C}{T_H}$$

Notice that for $T_C = 0$ K the efficiency is 100% and that efficiency becomes greater than 100% below 0 K. Since an efficiency greater than 100% violates the first law of thermodynamics, this implies that 0 K is the minimum possible temperature. In fact the lowest temperature ever obtained in a macroscopic system was 20 nK, which was achieved in 1995 at NIST. Subtracting the right hand side of Equation 5 from the middle portion and rearranging gives:

$$\frac{q_H}{T_H}\,0\,\frac{q_C}{T_C} @ 0$$

where the negative sign indicates heat ejected from the system. This relationship suggests the existence of a state function, *S*, defined by:

$$dS @ \frac{dq_{\text{rev}}}{T}$$

where the subscript indicates a reversible process. The change of this state function around any cycle is zero, as is necessary for any state function. This function corresponds to the entropy of the system, which was described previously. Rearranging Equation 6 gives a new definition for temperature in terms of entropy and heat:

$$T @ \frac{dq_{\text{rev}}}{dS}$$

For a system, where entropy *S*(*E*) is a function of its energy *E*, the temperature *T* is given by:

$$T^{01} @ \frac{d}{dE} S(E),$$

i.e. the reciprocal of the temperature is the rate of increase of entropy with respect to energy.

Definition from statistical mechanics

The previous section elaborated the historical derivation relating entropy and heat. A modern definition of temperature is given by statistical mechanics. It is defined in terms of the fundamental degrees of freedom of a system.

Generalized temperature from single particle statistics

It is possible to extend the definition of temperature even to systems of few particles, like in a quantum dot. The generalized temperature is obtained by considering time ensembles instead of configuration space ensembles given in statistical mechanics in the case of thermal and particle exchange between a small system of fermions with a single/double occupancy system. The finite quantum grand partition ensemble′ obtained under the hypothesis of ergodicity and orthodicity, allows to express the generalized temperature from the ratio of the average time of occupation $\hat{o}_1$ and $\hat{o}_2$ of the single/double occupancy system:

$$T \,@\, k^{01} ln2 \frac{w_2}{w_1} (E \circ E_F (1 \,.\, \frac{3}{2N}))$$

where E_F is the Fermi energy which tends to the ordinary temperature when N goes to infinity.

Negative temperature

On the empirical temperature scales, which are not referenced to absolute zero, a negative temperature is one below the zero-point of the scale used. For example, dry ice has a sublimation temperature of –78.5°C which is equivalent to –109.3°F. On the absolute Kelvin scale, however, this temperature is 194.6 K. On the absolute scale of thermodynamic temperature no material can exhibit a temperature smaller than or equal to 0K, both of which are forbidden by the third law of thermodynamics.

In the quantum mechanical description of electron and nuclear spin systems that have a limited number of possible states, and therefore a discrete upper limit of energy they can attain, it is possible to obtain a negative temperature, which is numerically indeed less than absolute zero. However, this is not the macroscopic temperature of the material, but instead the temperature of only very specific degrees of freedom, that are isolated from others and do not exchange energy by virtue of the equipartition theorem.

A negative temperature is experimentally achieved with suitable radio frequency techniques that cause a population inversion of spin states from the ground state. As the energy in the system increases upon population of the upper states, the entropy increases as well, as the system becomes less ordered, but attains a maximum value when the spins are evenly distributed among ground and excited states, after which it begins to decrease, once again achieving a state of higher order as the upper states begin to fill exclusively. At the point of maximum entropy, the temperature function shows the behavior of a singularity, because the slope of the entropy function decreases to zero at first and then turns negative.

Since temperature is the inverse of the derivative of the entropy, the temperature formally goes to infinity at this point, and switches to negative infinity as the slope turns negative. At energies higher than this point, the spin degree of freedom therefore exhibits formally a negative thermodynamic temperature. As the energy increases further by continued population of the excited state, the negative temperature approaches zero asymptotically. As the energy of the system increases in the population inversion, a system with a negative temperature is not colder than absolute zero, but rather it has a higher energy than at positive temperature, and may be said to be in fact hotter at negative temperatures. When brought into contact with a system at a positive temperature, energy will be transferred from the negative temperature regime to the positive temperature region.

PRESSURE BELTS

Since the time of Guericke, it has been known that the air has weight. Under ordinary conditions, a cubic foot of air weighs about an ounce and a quarter. As the air has weight, it must follow that, the atmosphere must press upon the surface of the earth, and the pressure at any point will depend upon the amount of air above it. It will be less on the top of a mountain than that at the foot. But even at the sea level the pressure varies from day to day. Small but distinct pressure differences remain from place to place. If 1,013 mb is taken as standard sea level pressure, readings higher than this will frequently be observed in middle latitudes, occasionally upto 1040 mb or higher. These pressures are designated as "High Pressures"; ranging down to 982 mb or below are "Low Pressures".

Units of Pressure: 1.013 kg/cm2 = 1.013 bar = 1,013 mb = 760 mm of Hg = 76 cm of Hg = 29.92 in.

EQUATORIAL LOW PRESSURE TROUGH

Within a few degrees of equator is a belt of pressure, some what lower than the normal pressure 1013 mb, between 1011 mb and 1008 mb, which is known as the Equatorial Trough or Low Pressure Belt. This is an area of high temperature and high humidity, commonly known as the Doldrums, where the air near sea level is stagnant or sluggish. The low pressure is due to heating; as the pressure of a volume of air decreases when its temperature increases. It lies entirely north of the equator during July, because this is the summer hemisphere.

The trough is by no means uniform in width, depth or position. It is deepest and most pronounced when it lies farthest from the equator, such as over an area extending from the Persian Gulf eastward to north-western India, and also over northern Mexico and the south-western United States. It is least distinct over the western and central pacific. Such continental, thermal lows are relatively shallow and have influence only on surface wind patterns.

SUB-TROPICAL HIGH PRESSURE BELTS

At about latitude 30 degrees North and South occur the sub-tropical high pressure belts, sometimes known as the "Horse Latitudes", zones of calm and descending air currents. In the Southern Hemisphere, this belt is clearly defined but contains centers of high pressure termed as "pressure cells". In the Northern Hemisphere in summer, the high pressure belt is dominated by two oceanic cells, one over the eastern Pacific and the other over the eastern north Atlantic. Average pressure exceeds 1026 mb in the centers of the cells. The sub-tropical highs are largely developed by dynamic, rather than thermal causes.

Subsidizing air from high levels is largely responsible for their great pressures. Variations in their strength and form are among the most important features of the global energy balance, since the air that diverges from them comprises a large part of the entire air circulation system on the surface. The summer hemisphere shows only the large oceanic, high pressure centers. The winter hemisphere has a greater number of anti-cyclonic highs and the most active day by day pressure changes along the general sub-tropical pressure zones.

SUB-ANTARCTIC LOW PRESSURE BELTS

Pole-wards of the sub-tropical high pressure belts are the broad belts of low pressure, extending roughly from the middle latitude zone to the Arctic zone but centered and intensified in the sub-Arctic zone at about 60th parallel to latitude. In the southern hemisphere, over the continuous expense of southern ocean, the sub-Antarctic low pressure belt is specially defined with average pressure as low as 984 mb.

It is one of the deepest and most persistent low pressure troughs in the world bordering the Antarctic continent. It is present, however, at all seasons shifting slightly southwards in January. This low pressure trough marks the zone of energy transfer between warm and cold air. It is a strong frontal zone, and the dynamics of air interchange, implemented by the upward displacement of warmer air, the high condensation, great angular momentum, rotational deflection, and other factors produce an almost continuous succession of deep cyclones that move around the world through this zone. Anticyclones, or high pressures accompanied by descending diverging air, sometimes, occupy positions within the trough and at rare intervals may persist for several days.

SUB-POLAR LOW PRESSURE BELTS

Nearer the poles occur the sub-polar low pressure belts. One reason for this pressure distribution is that the rotation of the earth causes a polar whirl and therefore a tendency toward low pressure at the poles. But the intense cold around the poles causes the thermal effect to overcome the dynamic one, with the result that the low pressure belts tend to be around just outside the polar circles. Since the frontal activity between cold and warm air in these latitudes is weak during the summer season, strong sub-polar lows do not develop. The weak low-pressure zone bordering southern Greenland and extending across the north Atlantic to Norway is related to shallow summer cyclonic depressions that pass eastward through the region.

POLAR HIGHS

The polar zones have permanent centers of high pressure known as polar highs. Both high and low pressure centers are present, which change in intensity and shape seasonally, with the low dominant during most of the year, but less strongly developed during the winter months. A high pressure ridge tends to occupy the highest portion of the continent but is extremely shallow and not well developed.

RELATION BETWEEN WIND AND PRESSURE

Wind may be defined as the movement of air from high to low atmospheric pressure in a virtually horizontal plane. Its direction and strength is the result of following four factors:

The Barometric or Pressure Gradient

Regions of high pressure form centers from which winds tend to blow outwards *i.e.*, they are areas of "divergence". Conversely, regions of low pressure are the face of winds *i.e.*, areas of "convergence". The change in barometric pressure across the horizontal surface constitutes a "pressure gradient", its direction is indicated by a broad arrow in the diagram. The gradient is in the direction from higher pressure to lower pressure. Where a pressure gradient exists, air molecules tend to drift in the same direction as that of the gradient.

This tendency for mass movement of the air is referred to as Pressure Gradient Force. The magnitude of the force is directly proportional to the steepness of the gradient. When the difference in pressure between two neighboring points on the earth's surface is great, there is said to be a steep pressure gradient. In such case, the resulting wind is strong. When, however, the barometric gradient is gentle, winds are light. The result is that the air tends to move from areas of high pressure to those of low. Wind is thus the horizontal motion of air in response to the pressure gradient force.

The Coriolis Force

If the earth did not rotate upon its axis, winds would have followed the direction of pressure gradient. Instead, earth's rotation produces another force, the Coriolis Force, which tends to turn the flow of air. The direction of action of the coriolis force is stated in Ferrel's Law:- "Any object or fluid moving horizontally in the Northern Hemisphere, tends to be deflected to the right of its path of motion, regardless of the compass direction of the path. In the Southern Hemisphere, a similar deflection is towards the left of the path of the motion as a result of the earth's rotation".

The coriolis force is absent at the equator but increases progressively polewards. The force is proportional to the speed of the moving object and it varies with latitude, being zero at the equator and maximum at the poles. Applying these principles to the relation of winds to pressure, the gradient force and the coriolis force quickly reach a balance or equilibrium when the wind has been turned to the point that it flows in a direction at right angles to the pressure gradient, *i.e.*, parallel with isobars. When a balanced condition develops between the forces exerted by the pressure gradient in one direction and the coriolis force in opposite direction, a steady wind will blow. This is known as a state of geostrophic balance, and the air movement as geostrophic flow or geostrophic wind, in a direction parallel to the straight isobars.

Centrifugal Force

When an air-stream moves on a curved course, as in a pressure system with closed or curved isobars, it is subjected to centrifugal force acting outwards from the center of curvature. A gradient wind develops as a result of the balance between the pressure gradient on the one hand, and the coriolis and centrifugal forces on the other.

The air thus tends to travel along the isobars, clockwise around the highs and anticlockwise around the lows in the Northern Hemisphere; while in Southern Hemisphere, it would tend to move anticlockwise around the highs and clockwise around the lows. This relation of the wind to atmospheric pressure was put forward by Dutch scientist Buys Ballot, in the middle of the 19th century:- "If you stand with your back to the wind in the Northern Hemisphere, pressure is lower on your left than on your right, and the reverse applies in the Southern Hemisphere".

Friction

Friction tends to decrease air velocities. Air that moves over the earth's surface has some of its mechanical energy transformed into other forms of energy because of friction, and thus its velocity decreases. Friction with surface, since it decreases wind velocity, also tends to decrease the amount of coriolis deflection, which is proportional to straight line velocity. This decrease in the coriolis deflection, in turn, tends to alter the resultant wind direction with respect to the pressure gradient.

The wind direction no longer parallels the isobars, as in geostrophic flow, but crosses them some what in the direction of pressure gradient. The angle of intersection is roughly proportional to the amount of friction encountered at the earth's surface. Surface winds usually cross isobars at about 10o angle over oceans and may reach angles of upto 45o over rough land surfaces. Although the loss of energy by friction takes place only at the surface, the breaking effect may be transmitted upwards in the higher levels of the horizontal wind flow. Usually, however, it is contained within 1500 to 2000 ft. of the surface.

MONSOON WINDS

Monsoons are the winds, which change their direction in different seasons. This is due to the fact that the surface of the earth, which consists of land and sea masses, acts differently to the insolation applied to it and accordingly the pressure is also different.

Cause of the Monsoons

The unequal heating of land and water areas causes monsoons. During summer, when the land has become quite hot, the contiguous ocean areas are relatively cooler. The air of the land rises up in the form of convection currents and the moist air of the oceans would flow in to take its place. During winter, the conditions would reverse and winds would blow from the lands towards oceans. In lower latitudes, the temperatures are uniformly high throughout the year while in the higher latitudes, the temperatures over the land are never so high as to disturb the prevailing pressure conditions.

Hence these types of circulation only take the form of land and sea breezes. But in the middle latitudes, because of vast range of temperature conditions, these land and sea breezes become semi-permanent and are known as the monsoon winds. These monsoons are the most powerful on the sub-continent of Indo-Pak and China.

Monsoons in Different Seasons

Summer

During summer, the interior of Asia and India becomes center of great heat and a powerful low pressure develops there. Comparatively the water surface of Indian Ocean is cooler and the pressure gradient is towards the heated interior. This factor is so important that trade winds are substituted by monsoon winds blowing from the south-west. Since it moves from water to land, the summer monsoon is moist and brings rain.

Winter

In the winter season, the interior of Asia and India is very cold and high pressure area exists there. In comparison, the contiguous areas of India Ocean are warmer and the pressure gradient exists towards the outlying ocean. As such the wind begins to blow from the interior of the continent towards the ocean and is known as the north-east monsoon. Coming from the dry interior, it is cold and dry. It only brings rain when it reaches some portion after crossing the sea.

Monsoons Winds of South & Southeast Asia

South Asian lands have mostly a tropical monsoon type of climate. They have a separate and distinct system of winds and air masses. In the extremely northern areas like Kashmir, Nepal and Bhutan, the winters are distinctly cold and wet, so they become a little different from the true monsoon climate while the eastern sides of south Asia *i.e.*, the ASEAN and Bangladesh areas may be called the true monsoon type. True monsoon types of climate lands experience fairly uniform distribution of temperatures and little diurnal range. Low pressures prevail in the summers, which result in the abundance of rainfall by summer monsoons and long drought periods in winter season for more than four months. So summers are quite hot and rainy, while winters in the true monsoon climate are almost dry and warm. The summer monsoon comes to South Asia from the Indian Ocean as south-western winds, then it curves towards the north-west upto Ganges plain and then towards the Indus plain. Usually monsoon winds of South Asia are sub-divided into two parts:

- The period from June to mid September, a time of advancing monsoon winds.
- The period from mid September to mid December, a time retreating monsoon winds.

During the hot dry season, due to very high temperatures in the desert and semi-desert areas of South Asia, very low pressure areas are created, which eventually cause the outburst of the monsoon. This low pressure area attracts winds from Indian Ocean, full of humidity, giving heavy rains over the South Asian region. The duration of this monsoon varies from six months in the extreme south of Madras, to only six weeks in the Pakistan area. The speed of monsoon winds also varies in different parts of South Asia. There are two main routes of these monsoon winds. One route starts from Arabian Sea and scarcely reaches the north of the gulf of Cambay. The second route starts from the Bay of Bengal as an east wind, upto the Ganges and reaches Punjab from the south-east.

In the eastern side, *i.e.*, Bengal, the monsoon starts from the middle of June and reaches Punjab in the month of July. South-East Asia receives continuous and the heaviest rainfalls as compared to the others parts of South Asia. Rainfall reduces the temperatures in most parts of India and Pakistan. The areas of Punjab and Sindh are the driest parts in this season and have the highest temperatures, which remain in June and July. In northern plain areas, the rainfall is from 30 in. to 60 in.

on the leeward side of mountains. In the interior of plateau and valleys of peninsular India, the rainfall received is only 10 in. to 30 in. When this summer monsoon comes towards the west, rainfall gradually decreases because dry hot air may reach across central India. No rainfall occurs here. Only the Himalayan regions receive heavy rainfall during this time. Kashmir receives 20 in. to 40 in. of rainfall. The Rajisthan and Sindh areas receive no rain during this rainy period. The winter monsoon of South Asia is dry with cool air travelling across the Indo-Gangetic Plan from the north-west, and in part, from subsidiary air in the sub-tropical high pressure belt. The winter monsoon moves from west to east and then turns to become north-easterly over the peninsular India. The mean January temperature remains above 70 oF. The rainfall is less than 5 to 13 cm in Punjab and more than 5 to 13 cm in Himalayan areas with cold air blowing.

Monsoon Winds of North America

North America does not have the remarkable extremes of monsoon winds experienced by South-Eastern Asia, but there is nevertheless an alternation of temperature and pressure conditions between winter and summer. Wind records show that in summer, there is a prevailing tendency for air originating in the gulf of Mexico to move northward across the central and eastern parts of the United States; whereas in winter, there is a prevailing tendency for air to move southward from sources in Canada. Australia, too, shows a monsoon effect, but being south of the equator, it reverses the conditions of Asia.

EARTH'S SURFACE WIND SYSTEM

When the movement of the air in the atmosphere is in a horizontal direction over the surface of the earth, it is known as the wind. Movement of the wind is directly controlled by pressure.

The Equatorial Belt of Variable Winds

Over the equatorial trough of low pressure, lying roughly between 5o S and 5o N latitude, is the equatorial belt of variable winds and calms or the Doldrums. There are no surface winds here, but a fair distribution of directions around the compass. Calms prevail as much as a third of the time. Centrally located on a belt of low pressure, this zone has no strong pressure gradients to induce a persistent flow of wind.

Trade Wind Belt

North and south of the doldrums are the trade wind belts, covering the roughly the zones lying between 5o and 30o north and south. The trade is a result of a pressure gradient from sub-tropical belts of high pressure to the equatorial trough of low pressure. In the northern hemisphere, air moving equator-wards is deflected by the earth's rotation to turn westwards. Thus, the prevailing wind is from the north-east and winds are termed as the north-east trades. In the Southern Hemisphere, deflection of the moving air to the left causes the south-east trades.

Trade winds are noted for their steadiness and directional persistence. Most winds come from one quarter of the equator. The trades are best developed over the pacific and Atlantic oceans, but are upset in the India Ocean region by the proximity of great Asiatic land mass. The trade wind belts are not altogether favorable for navigation or flying, because over certain oceanic portions, at certain seasons of the year, terrible tropical storms known as hurricanes or typhoons occur. Sub-Tropical Belt of Variable

Winds and Calms

Between latitudes 30o and 40o north and south are what has long been called the sub-tropical belt of variable winds and calms or horse latitudes, coinciding with the sub-tropical high pressure belt. Instead of being continuos even belts, the high pressure areas are concentrated into distinct anticyclones or cells, located over the oceans. The apparent outward spiraling movement of air is directed equatorward into easterly trade wind system; polewards into the westerly wind system. The cells of higher pressure are most strongly developed in summer.

There is also a latitudinal shifting following the sun's declination. This amounts to less than 5o in the Southern Hemisphere, but is about 8o for the strong Hawaiian high latitude in the north-eastern Pacific. Winds in the high pressure cells are distributed around a considerable range of compass directions. Calms prevail as much as a quarter of the time. The cells have generally fair, clear weather, with a strong tendency to dryness. Most of the world's great deserts lie is this zone and in the adjacent trade wind belt.

Belt of Westerlies

Between latitudes 40o and 60o north and south, is the belt of westerlies or prevailing westerly winds. Within the westerly wind belt, winds blow from any direction of the compass, but the westerly components are definitely predominant. Storm winds are common in this belt, as are frequent cloudy days with continued precipitation. Weather is highly changeable. In the Northern Hemisphere, land masses cause considerable disruption of the westerly wind belt, but in the Southern Hemisphere, between the latitude 40o and 60o south, there is an almost unbroken belt of ocean. Here the westerlies gain great strength and persistence.

The belt was extensively used for selling vessels travelling eastwards from the south Atlantic Ocean to Australia, Tasmania, New Zealand and the Southern Pacific Islands. Although the westerly wind belts no longer exert a strong influence over the routes of modern ocean vessels, they are important in long distance flying. Transoceanic and transcontinental flights in the easterly direction require less fuel and shorter time. On westward flights, strong headwinds may eat dangerously into the fuel supply on the plane and in any event necessitate reduced payload.

Polar Easterlies

A wind system termed polar easterlies has been described as the characteristic of the Arctic and the polar zones. The concept is greatly over simplified, if not actually erroneous, for winds in these regions take a variety of directions, as dictated by local weather disturbances. Perhaps in Antarctica, where an ice-capped land mass rests squarely upon the pole and is surrounded by a vast oceanic expense, the outward spiraling flow of polar easterlies is a valid concept. Deflected to the left in the Southern Hemisphere, the radial winds would spiral counter clockwise, producing a system of south-easterly winds.

HEAT BUDGET OF THE EARTH'S

The Earth's Heat Budget is a description of what happens to the radiation Earth receives from the sun. The Earth needs this radiation to drive forces such as the circulation of the atmosphere and the ocean. This energy and the way the atmosphere and land respond to is, gives us our weather and climate. In the summer months, June-August, the northern hemisphere receives more radiation than the southern hemisphere because the Earth is tilted on its axis. The equator receives more energy throughout the whole year than the poles because the sun's rays do not strike at an angle.

The atmosphere interferes with solar radiation in two ways: scattering and absorption. All molecules tend to scatter radiation, the character of the scattering depends on the wavelength of the radiation relative to the size the molecules or other particles in the atmosphere.

Our atmosphere is made up of very small molecules relative to wavelengths of the sunlight, and because of this, shorter wavelength radiation is scattered more effectively than longer wavelengths through a process known as Rayleigh scattering. This is the reason for the sky being blue. The many tiny molecules in the Earth's atmosphere favour the shorter wavelengths, such as blue light. Gases and aerosols absorb some solar radiation. It is usually relatively small and less important than scattering. However, ultraviolet radiation is strongly absorbed by ozone, which results in shielding the surface from UV radiation.

Of all the radiation that is received by Earth's atmosphere:

- 31% is reflected back into space by the atmosphere
- 4% is reflected back by the Earth's surface
- 59.5% is re-radiated from the Earth's atmosphere back into space
- 5.5% is re-radiated from the Earth's surface back into space

Reflection means that the sun's light 'bounces' off the surface, not changing the wave length. Re-radiation means that the radiation is absorbed by the material and heats it. The mean global surface temperature of the earth is 15 degrees Celsius, however because much of the Earth's atmosphere is cooler than the surface, Earth emits radiation to space equivalent of a body with a temperature of -18 degrees Celsius and it is long wave radiation. Much of this is trapped in the atmosphere by so-called greenhouse gases, heating the atmosphere- producing the Greenhouse Effect. The proportion of this radiation that is reflected outward is known as the surface albedo. Some examples of the surface albedo values for Earth are:

- Tropical forest- 13%
- Woodland- 14%
- Farmland/natural grassland- 20%
- Semi-desert/stony desert- 24%
- Dry sandy desert/salt plans- 37%
- Water (0-60 degrees) - &lt8%
- Water (60-90 degrees) - &lt10%
- Sea ice – 25-60%
- Snow-covered vegetation – 20-80%
- Snow-covered ice – 80%

Most of the radiation that hits the Earth's surface is absorbed by dark soil and rock. The oceans also absorb a lot of the energy it receives, Water has a high heat capacity, which means that even though it absorbs large amounts of heat, its temperature will only rise a little. In this way, the Earth keeps itself at equilibrium, by not warming up or cooling down.

ATMOSPHERIC CIRCULATION

To understand large-scale motions of the atmosphere, it is essential that the Aerographer's Mate study the general circulation of the atmosphere. The sun's radiation is the energy that sets the atmosphere in motion, both horizontally and vertically. The rising and expanding of the air when it is warmed, or the descending and contracting of the air when it is cooled causes the vertical motion. The horizontal motion is caused by differences of atmospheric pressure; air moves from areas of high pressure toward areas of low pressure.

Differences of temperature, the cause of the pressure differences, are due to the unequal absorption of the Sun's radiation by Earth's surface. The differences in the type of surface; the differential heating; the unequal distribution of land and water; the relative position of oceans to land, forests to mountains, lakes to surrounding land, and the like, cause different types of circulation of the air. Due to the relative position of Earth with respect to the Sun, much more radiation is absorbed near the equator than at other areas, with the least radiation being absorbed at or near the poles. Consequently, the principal factor affecting the atmosphere is incoming solar radiation, and its distribution depends on the latitude and the season.

GENERAL CIRCULATION

The general circulation theory attempts to explain the global circulation of the atmosphere with some minor exceptions. Since Earth heats unequally, the heat is carried away from the hot area to a cooler one as a result of the operation of physical laws. This global movement of air, which restores a balance of heat on Earth, is the general circulation.

WORLD TEMPERATURE GRADIENT

Temperature gradient is the rate of change of temperature with distance in any given direction at any point. World temperature gradient refers to the change in temperature that exists in the atmosphere from the equator to the poles. The change in temperature or temperature differential, which causes atmospheric circulation can be compared to the temperature differences produced in a pan of water placed over a gas burner.

As the water is heated, it expands and its density is lowered. This reduction in density causes the warmer, less dense water to rise to the top of the pan. As it rises, it cools and is forced to the edges of the pan. Here it cools further and then sinks to the bottom, eventually working its way back to the center of the pan where it started. This process sets up a simple circulation pattern due to successive heating and cooling.

Ideally, the air within the troposphere may be compared to the water in the pan. The most direct rays of the Sun hit Earth near the equator and cause a net gain of heat. The air at the equator heats, rises, and flows in the upper atmosphere toward both poles. Upon reaching the poles, it cools sufficiently and sinks back toward Earth, where it tends to flow along the surface of Earth back to the equator.

Simple circulation of the atmosphere would occur as described above if it were not for the following factors:

- Earth rotates, resulting in an apparent force known as the Coriolis force. This rotation results in a constant change to the area being heated.
- Earth is covered by irregular land and water surfaces that heat at different rates.

Regions under the direct rays of the Sun absorb more heat per unit time than those areas receiving oblique rays. The heat produced by the slanting rays of the Sun during early morning may be compared with the heat that is produced by the slanting rays of the Sun during winter. The heat produced by the more direct rays at midday can be compared with the heat resulting from the more direct rays of summer.

The length of day, like the angle of the Sun's rays, influences the temperature. The length of day varies with the latitude and the season. Near the equator there are about 12 hours of daylight with the Sun's rays striking the surface more directly. Consequently, equatorial regions normally do not have pronounced seasonal temperature variations.

During the summer in the Northern Hemisphere, all areas north of the equator have more than 12 hours of daylight. During the winter the situation is reversed; latitudes north of the equator have less than 12 hours of daylight. Large seasonal variation in the length of the day and the seasonal difference in the angle at which the Sun's rays reach Earth's surface cause seasonal temperature differences in middle and high latitudes.

ELEMENTS OF CIRCULATION

Temperature differences cause pressure differences, which in turn cause air movements. The following sections show how air movements work and how they evolve into the various circulations—primary, secondary, and tertiary.

To explain the observed wind circulation over Earth, three basic steps are used. The first step is to assume Earth does not rotate and is of uniform surface; that is, all land or all water. The second step is to rotate Earth, but still assume a uniform surface. The third step is to rotate Earth and assume a non-uniform surface. For now, we deal with the first two steps, a non-rotating Earth of uniform surface and a rotating Earth of uniform surface.

Static Earth

The circulation on a non-rotating Earth is referred to as the thermal circulation because it is caused by the difference in heating. The air over the equator is heated and rises; while over the poles the air is cooled and sinks.

Rotating Earth

In thermal circulation, the assumption was made that the Earth did not rotate, but of course this is not true. The rotation of Earth causes a force that affects thermal circulation. This rotation results in the deflection to the *right* of movement in the Northern Hemisphere, and to the *left* of the movement in the Southern Hemisphere. This force is called the *Coriolis* force. The Coriolis force is not a true force. It is an apparent force resulting from the west-to-east rotation of Earth. The effects, however, are real.

Arctic rivers cut faster into their right banks than their left ones. On railroads carrying only one-way traffic, the right hand rails wear out faster than the left-hand rails. Artillery projectiles must be aimed to the left of target because they deflect to the right. Pendulum clocks run faster in high latitudes than in lower latitudes. All these phenomena are the result of the Coriolis force, which is only an apparent force.

The most important phenomena are that this force also deflects winds to the right in the Northern Hemisphere. Therefore, it is important to understand how this force is produced. As Earth rotates, points on the surface are moving eastward past a fixed point in space at a given speed. Points on the equator are moving at approximately 1,000 miles per hour, points on the poles are not moving at all, but are merely pivoting, the points somewhere between are moving at speeds between 1,000 and zero miles per hour depending upon their relative position.

Assume that a missile located at the North Pole is launched at a target on the equator. The missile does not have any eastward lateral velocity, but the target has an eastward velocity of 1,000 miles per hour. The result is that the missile appears to be deflected to the right as the target moves away from its initial position.

A similar condition assumes that a missile located on the equator is launched at a target at the North Pole. The missile has an eastward lateral velocity of 1,000 miles per hour, while the target on the pole has no lateral velocity at all. Once again the missile appears to be deflected to the right as a result of its initial eastward lateral velocity. Due to Earth's rotation and the Coriolis effect, the simple circulation now becomes more complex. The complex on resulting from the interplay of the Coriolis effect with the flow of air is known as the theory.

3-CELL THEORY

According to the 3-cell theory, Earth is divided into six circulation belts—three in the Northern Hemisphere and three in the Southern Hemisphere. The dividing lines are the equator, latitude, and 60°N and S latitude. The general circulation of the Northern Hemisphere is similar to those of the Southern Hemisphere. First, note the tropical cell of the Northern Hemisphere that lies between the equator and 30°N latitude.

Convection at the equator causes the air to heat and rise, due to convection. When it reaches the upper portions of the troposphere, it tends to flow toward the North Pole. By the time the air has reached 30°N latitude, the Coriolis effect has deflected it so much that it is moving eastward instead of northward. This results in a piling up of air near 30°N latitude and a descending current of air toward the surface which forms a belt of high pressure. When the descending air reaches the surface where it flows outward, part of it flows poleward to become part of the mid-latitude cell; the other part flows toward the equator, where it is deflected by the Coriolis effect and forms the northeast trades.

The mid-latitude cell is located between 30° and 60°N latitude. The air, which comprises this cell, circulates poleward at the surface and equatorward aloft with rising currents at 60° and descending currents at 300. However, in general, winds both at the surface and aloft blow from the west. The Coriolis effect easily explains this for the surface wind on the poleward-moving surface air. The west wind aloft is not as easily explained. Most authorities agree that this wind is frictionally driven by the west winds in the two adjacent cells. The polar cell lies between 60°N latitude and the North Pole. The circulation in this cell begins with a flow of air at a high altitude toward the pole. This flow cools and descends at the North Pole and forms a high-pressure area in the Polar Regions. After reaching the surface of Earth, this air usually flows equatorward and is deflected by the Coriolis effect so that it moves from the northeast. This air converges with the poleward flow from the mid-latitude cell and is deflected upward with a portion circulating poleward again and the remainder equatorward.

The outflow of air aloft between the polar and mid-latitude cells causes a semi-permanent low-pressure area at approximately 60°N latitude. To complete the picture of the world's general atmospheric circulation, we must associate this prevailing wind and pressure belts with some basic characteristics.

WORLD WINDS

In the vicinity of the equator is a belt of light and variable winds known as the doldrums. On the poleward side of the doldrums are the *trade winds*; the predominant wind system of the tropics. These easterly winds are the most consistent on Earth, especially over the oceans. Near 30°N and 30°S latitudes lie the sub-tropical high-pressure belts. Winds are light and variable. These areas are referred to as the *horse latitudes*.

The prevailing westerlies, which are on the poleward side of the subtropical high-pressure belt, are persistent throughout the mid-latitudes. In the Northern Hemisphere, the direction of the westerlies at the surface is from the southwest. In the Southern Hemisphere, westerlies are from the northwest. This is due to the deflection area resulting from the Coriolis effect as the air moves poleward. Poleward of the prevailing westerlies, near 60°N and 60°S latitudes, lies the belt of low-pressure basic pressure known as the *polar front zone*. Here, converging winds result in ascending air currents and consequent poor weather.

WIND THEORY

Newton's first two laws of motion indicate that motion tends to be in straight lines and only deviates from such lines when acted upon by another force or by a combination of forces. Air tends to move in a straight line from a high-pressure area to a low-pressure area. However, there are forces that prevent the air from moving in a straight line.

Wind Forces

There are four basic forces that affect the directional movement of air in our atmosphere: pressure gradient force, the Coriolis effect, centrifugal force, and frictional force. These forces, working together, affect air movement. The forces that are affecting it at that particular time determine the direction that the air moves. Also, the different names given to the movement of the air depends on what forces are affecting it.

Pressure Gradient

The rate of change in pressure in a direction perpendicular to the isobars is called *pressure gradient*. Pressure applied to a fluid is exerted equally in all directions throughout the fluid; *e.g.*, if a pressure of 1013.2 millibars is exerted downward by the atmosphere at the surface, this same pressure is also exerted horizontally outward at the surface. Therefore, a pressure gradient exists in the horizontal as well as the vertical plane in the atmosphere.

Horizontal Pressure Gradient

The horizontal pressure gradient is *steep* or *strong* when the isobars determining the pressure system are close together. It is *flat* or *weak* when the isobars are far apart.

Vertical Pressure Gradient

If isobars are considered as depicting atmospheric topography, a high-pressure system represents a hill of air, and a low-pressure system represents a depression or valley of air. The vertical pressure gradient always indicates a decrease in pressure with altitude, but the rate of pressure decrease varies directly with changes in air density with altitude. Below 10,000 feet altitude, pressure decreases approximately 1 inch of mercury per 1,000 feet in the standard atmosphere. The vertical cross section through a high and low depicts the vertical pressure gradient.

Pressure Gradient Force

The variation of heating from one locality to another is the initial factor that produces movement of air or wind. The most direct path from high to low pressure is the path along which the pressure is changing most rapidly. The rate of change is called the pressure gradient. Pressure gradient force is the force that moves air from an area of high pressure to an area of low pressure. The velocity of the wind depends upon the pressure gradient. If the pressure gradient is strong, the wind speed is high. If the pressure gradient is weak, the wind speed is light. Instead the flow is circular around the pressure systems. Pressure gradient force causes the air to begin moving from the high-pressure to the low-pressure system. Coriolis force and centrifugal force then begin acting on the flow in varying degrees. In this example, frictional force is not a factor.

Coriolis Effect

If pressure gradient force were the only force affecting windflow, the wind would blow at right angles across isobars from high to low pressure. The wind actually blows parallel to isobars above any frictional level. Therefore, other factors must be affecting the windflow; one of these factors is the rotation of

Earth. A particle at rest on Earth's surface is in equilibrium. If the particle starts to move because of a pressure gradient force, its relative motion is affected by the rotation of Earth. If a mass of air from the equator moves northward, it is deflected to the right, so that a south wind tends to become a southwesterly wind.

In the Northern Hemisphere, the result of the Coriolis effect is that moving air is deflected to the right of its path of motion. This deflection to the right is directly proportional to the speed of the wind; the faster the wind speed, the greater the deflection to the right, and conversely, the slower the wind speed, the less the deflection to the right. Finally, this apparent deflective force is stronger at the Polar Regions than at the equator.

Centrifugal Force

According to Newton's first law of motion, a body in motion continues in the same direction in a straight line and with the same speed unless acted upon by some external force. Therefore, for a body to move in a curved path, some force must be continually applied. The force restraining bodies that move in a curved path is called the *centripetal force*; it is always directed *toward* the center of rotation. When a rock is whirled around on a string, the centripetal force is afforded by the tension of the string.

Newton's third law states that for every action there is an equal and opposite reaction. *Centrifugal force* is the reacting force that is equal to and opposite in direction to the centripetal force. Centrifugal force, then, is a force directed *outward* from the center of rotation.

As you know, a bucket of water can be swung over your head at a rate of speed that allows the water to remain in the bucket. This is an example of both centrifugal and centripetal force. The water is held in the bucket by centrifugal force tending to pull it outward. The centripetal force, the force holding the bucket and water to the center, is your arm swinging the bucket. As soon as you cease swinging the bucket, the forces cease and thewater falls out of the bucket.

High- and low-pressure systems can be compared to rotating discs. Centrifugal effect tends to fling air out from the center of rotation of these systems. This force is directly proportional to the wind speeds, the faster the wind, and the stronger the outward force. Therefore, when winds tend to blow in a circular path, centrifugal effect influences these winds.

Frictional Force

The actual drag or slowing of air particles in contact with a solid surface is called friction. Friction tends to retard air movement. Since Coriolis force varies with the speed of the wind, a reduction in the wind speed by friction means a reduction of the Coriolis force. This results in a momentary disruption of the balance. When the new balance is reached, the air flows at an angle across the isobars from high pressure to low pressure. This angle varies from 10 degrees over the ocean to more than 45 degrees over rugged terrain. Frictional effects on the air are greatest near the ground, but the effects are also carried aloft by turbulence. Surface friction is effective in slowing the wind to an average altitude of 2,000 feet above the ground. Above this level, called the gradient wind level or the second standard level the effect of friction decreases rapidly and may be considered negligible. Air above 2,000 feet normally flows parallel to the isobars.

WIND TYPES

Since there is a direct relationship between pressure gradient and wind speed and direction, we have a variety of wind types to deal with. We discuss below the relationship of winds and circulations, the forces involved, and the effect of these factors on the general circulation.

Geostrophic and Gradient Wind

On analyzed surface weather charts, points of equal pressure are connected by drawn lines referred to as isobars, while in upper air analysis, points of equal heights are connected and called isoheights. The variation of these heights and pressures from one locality to another is the initial factor that produces movement of air, or wind. Assume that at three stations the pressure is lower at each successive point. This means that there is a horizontal pressure gradient for each unit distance. With this situation, the air moves from the area of greater pressure to the area of lesser pressure.

If the force of the pressure were the only factor acting on the wind, the wind would flow from high to low pressure, perpendicular to the isobars. Since experience shows the wind does not flow perpendicular to isobars, but at a slight angle across them and towards the lower pressure, it is evident that other factors are involved. These other factors are the Coriolis effect, frictional force, and centrifugal effect. When a unit of air moves with no frictional force involved, the movement of air is parallel to the isobars. This wind is called a gradient wind.

When the isobars are straight, so only Coriolis and pressure gradient forces are involved, it is termed a geostrophic wind. Let's consider a parcel of air from the time it begins to move until it develops into a geostrophic wind. As soon as a parcel of air starts to move due to the pressure gradient force, the Coriolis force begins to deflect it from the direction of the pressure gradient force.. The Coriolis force is the apparent force exerted upon the parcel of air due to the rotation of Earth. This force acts to the right of the path of motion of the air parcel in the Northern Hemisphere. It always acts at right angles to the direction of motion. In the absence of friction, the Coriolis force changes the direction of motion of the parcel until the Coriolis force and the pressure gradient force are in balance. When the two forces are equal and opposite, the wind blows parallel to the straight isobars. The Coriolis force only affects the direction, not the speed of the motion of the air. Normally, Coriolis force is not greater than the pressure gradient force. In the case of super-gradient winds, Coriolis force may be greater than the pressure gradient force. This causes the wind to deflect *more* to the right in the Northern Hemisphere, or toward higher pressure. Under actual conditions, air moves around high and low pressure centers toward lower pressure. Here, the flow of air is from the area of high pressure to the area of low pressure, but, as we mentioned previously, it does not flow straight across the isobars. Instead, the flow is circular around the pressure systems.

The Coriolis force commences deflecting the path of movement to the right or left until it reaches a point where a balance exists between the Coriolis and the pressure gradient force. At this point the air is no longer deflected and moves forward around the systems. Once circular motion around the systems is established, then centrifugal force must be considered. Centrifugal force acts outward from the center of both the highs and the lows with a force dependent upon the velocity of the wind and the degree of curvature of the isobars. However, the pressure gradient force is acting towards the low; therefore, the flow in that direction persists. When it is moving parallel to that portion of the analysis showing straight flow, it is a *geostrophic wind*.

We defined pressure gradient as being a change of pressure with distance. This means that if the isobars are closely spaced, then the pressure change is greater over a given distance; it is smaller if they are widely spaced. Therefore, the closer the isobars, the faster the flow. Geostrophic and gradient winds are also dependent, to a certain extent, upon the density of the atmosphere and the latitude. If the density and the pressure gradient remain constant and the latitude increases, the wind speed decreases. If the latitude decreases, the wind speed increases. If the density and the latitude remain constant and the pressure gradient decreases, the wind speed decreases. If the pressure gradient and the latitude remain constant and the density decreases, the wind speed increases. If the density increases, the wind speed decreases. True geostrophic wind is seldom observed in nature, but the conditions are closely approximated on upper-level charts.

Cyclostrophic Wind

In some atmospheric conditions, the radius of rotation becomes so small that the centrifugal force becomes quite strong in comparison with the Coriolis force. This is particularly true in lowlatitudes where the Coriolis force is quite small to begin with. In this case, the pressure gradient force is nearly balanced by the centrifugal force alone. When this occurs, the wind is said to be cyclostrophic. By definition, a cyclostrophic wind exists when the pressure gradient force is balanced by the centrifugal force alone. This exact situation rarely exists, but is so nearly reached in some situations that the small Coriolis effect is neglected and the flow is said to be cyclostrophic. Winds in a hurricane or typhoon and the winds around a tornado are considered cyclostrophic.

Movement of Wind around Anticyclones

The movement of gradient winds around anticyclones is affected in a certain manner by the pressure gradient force, the centrifugal force, and the Coriolis force. The pressure gradient force acts from high to low pressure, and the Coriolis force acts opposite to the pressure gradient force and at right angles to the direction of movement of the parcel of air.

The centrifugal force acts at right angles to the path of motion and outward from the center about which the parcel is moving. In the case of a high-pressure center, the pressure gradient force and the centrifugal force balance the Coriolis force. This phenomenon may be expressed in the following manner:

$$PG + CF = D$$

Movement of Wind around Cyclones

As in the case of anticyclones, the pressure gradient force, the centrifugal force, and the Coriolis force affect gradient winds around cyclones, but the balance of the forces is different.. In a cyclonic situation the Coriolis force and the centrifugal force balance the pressure gradient force. This balance may be expressed in the following manner:

$$PG = D + CF$$

Centrifugal force acts *with* the pressure gradient force when the circulation is anticyclonic and *against* the pressure gradient force when the circulation is cyclonic. Therefore, wind velocity is greater in an anticyclone than in a cyclone of the same isobaric spacing.

Variations

It has been determined that, given the same density, pressure gradient, and latitude, the wind is weaker around a low-pressure cell than a high-pressure cell. This is also true for gradient and geostrophic winds. The wind we observe on a synoptic chart is usually stronger around low cells than high cells because the pressure gradient is usually stronger around the low-pressure cell.

Geostrophic and Gradient Wind Scales

The geostrophic wind is stronger than the gradient wind around a low and is weaker than a gradient wind around a high. This is why the isobar spacing and contour spacing, for a curved flow, differs from that determined by a geostrophic wind scale. If the flow under consideration is around a high-pressure cell, the isobars are farther apart than indicated by the geostrophic wind scale. If the flow is around a low-pressure cell, the isobars are closer together than indicated by the geostrophic wind scale. Geostrophic and gradient wind scales are used to determine the magnitude of these winds and to determine the isobar or contour spacing. There are a number of scales available for measuring geostrophic and gradient flowof both surface and upper air charts.

Weather plotting charts used by the Naval Oceanography Command has geostrophic wind scales printed on them for both isobaric and contour spacing. The most common scales in general use can be used for both surface and upper air charts. The scales are in 4mb and 60m intervals.

SECONDARY CIRCULATION

Now that you have a picture of the *general* circulation of the atmosphere over Earth, the next step is to see how land and water areas offset the general circulation. The circulations caused by the effect of Earth's surfaces, its composition and contour, are known as *secondary circulations*. These secondary circulations give rise to winds that often cancel out the normal effect of the great wind systems. There are two factors that cause the pressure belts of the primary circulation to break up into closed circulations of the secondary circulations. They are the non-uniform surface of the earth and the difference between heating and cooling of land and water. The surface temperature of oceans changes very little during the year. However, land areas sometimes undergo extreme temperature changes with the seasons. In the winter, large high-pressure areas form over the cold land and the low-pressure areas form over the relatively warm oceans. *The reverse is true in summer when highs are over water and lows form over the warm land areas. The result of this difference in heating and cooling of land and water surfaces is known as the thermal effect.*

Circulation systems are also created by the interaction of wind belts of pressure systems or the variation in wind in combination with certain distributions of temperature and/or moisture. This is known as the dynamic effect. This effect rarely, if ever, operates alone in creating secondary systems, as most of the systems are both created and maintained by a combination of the thermal and dynamic effects.

CENTERS OF ACTION

The pressure belts of the general circulation are rarely continuous. They are broken up into detached areas of high and low pressure cells by the secondary circulation. The breaks correspond with regions showing differences in temperature from land to water surfaces.

There is a permanent belt of relatively low pressure along the equator and another deeper belt of low-pressure paralleling the coast of the Antarctic Continent. Permanent belts of high pressure largely encircle Earth, generally over the oceans in both the Northern and Southern Hemispheres. The number of centers of action is at a maximum at about 30 to 35 degrees from the equator. There are also regions where the pressure is predominantly low or high at certain seasons, but not throughout the year. In the vicinity of Iceland, pressure is low most of the time. The water surface is warmer than the surface of Iceland or the icecaps of Greenland. The Icelandic low is most intense in winter, when the greatest temperature contrast occurs, but it persists with less intensity through the summer.

Near Alaska, a similar situation exists with the Aleutian low. The Aleutian low is most pronounced when the neighboring areas of Alaska and Siberia are snow covered and colder than the adjacent ocean. These lows are not a continuation of one and the same cyclone. They are, however, regions of low pressure where lows frequently form or arrive from other regions. Here they remain stationary or move sluggishly for a time, then the lows move on or die out and are replaced by others. Occasionally these regions of low pressure are invaded by traveling high-pressure systems.

Two areas of semi permanent high-pressure also exist. There is a semi permanent high-pressure center over the Pacific westward of California and another over the Atlantic, near the Azores and of the coast of Africa. Pressure is also high, but less persistently so, west of the Azores to the vicinity of Bermuda. These subtropical highs are more intense and cover a greater area in summer thàn winter. They also extend farther northward summer. In winter, these systems move soul toward the equator, following the solar equator. The largest individual circulation cells in the Northern Hemisphere are the Asiatic high in winter and the Asiatic low in summer. In winter, the Asiatic continent is a region of strong cooling and therefore is dominated by a large high-pressure cell.

In summer, strong heating is present and the high-pressure cell becomes a large low-pressure cell. This seasonal change in pressure cells gives rise to the monsoon flowover India and Southeast Asia. Another cell that is often considered to be a center of action is the polar high. Both Arctic and Antarctic highs have considerable variations in pressure, and these regions have many traveling disturbances in summer. For example, the Greenland high is a persistent feature, but it is not a well-defined high during all seasons of the year. The Greenland high often appears to be an extension of the polar high or vice versa.

Other continental regions showseasonal variations, but are generally of small size and their location is variable. Therefore, they are not considered to be centers of action. An average annual pressure distribution chart reveals several important characteristics. First, along the equator there is a belt of relatively low pressure encircling the globe with barometric pressure of about 1,012 millibars. Second, on either side of this belt of low pressure is a belt of high pressure. This high-pressure area in the Northern Hemisphere lies mostly between latitudes 30° and 40°N with three well-defined centers of maximum pressure. One is over the eastern Pacific, the second over the Azores and the third over Siberia; all are about 1,020 millibars. The belt of high pressure in the Southern Hemisphere is roughly parallel to 30°S. It, too, has three centers of maximum pressure. One is in the eastern Pacific, the second in the eastern Atlantic, and the third in the Indian Ocean; again, all are about 1,020 millibars. A third characteristic to be noted from this chart is that, beyond the belt of high pressure in either hemisphere, the pressure diminishes toward the poles. In the Southern Hemisphere, the decrease in pressure toward the South Pole is regular and very marked. The pressure decreases from, an average slightly above 1,016 millibars along latitude 35°S to an average of 992 millibars along latitude 60°S In the Northern Hemisphere, however, the decrease in pressure toward the North Pole is less regular and not as great.

This is largely due to the distribution of land and water: note the extensive landmass in the Northern Hemisphere as compared to those of the Southern Hemisphere. While the pressure belts that stand out on the average annual pressure distribution chart represent average pressure distribution for the year, these belts are rarely continuous on any given day. They are usually broken up into detached areas of high or low pressure by the secondary circulation of the atmosphere.

In either hemisphere, the pressure over the land during the winter season is decidedly above the annual average. During the summer season, the pressure is decidedly below the average, with extreme variations occurring such as in the case of continental Asia. Here the mean monthly pressure ranges from about 1,033 millibars during January to about 999 millibars during July. Over the northern oceans, on the other hand, conditions are reversed; the summer pressure there is somewhat higher. Thus in January the Icelandic and Aleutian lows intensify to a depth of about 999 millibars, while in July these lows fill and are almost obliterated.

The polar high in winter is not a cell centered directly over the North Pole, but appears to be an extension of the Asiatic high and often appears as a wedge extending from the Asiatic continent. The cell is displaced toward the area of coldest temperatures—the Asiatic continent. In summer, this high appears as an extension of the Pacific high and is again displaced toward the area of coolest temperature, which in this case is the extensive water area of the Pacific. In winter over North America, the most significant feature is the domination by a high-pressure cell. This cell is also due to cooling but is not as intense as the Asiatic cell. In summer, the most significant feature is the so-called heat low over the southwestern part of the continent, which is caused by extreme heating in this region.

MIGRATORY SYSTEMS

General circulation, based on an average of wind conditions, is a more or less quasi-stationary circulation. Likewise, much of the secondary circulation depends on more or less static conditions that, in turn, depend on permanent and semi permanent high and low-pressure areas. Changes in the circulation patterns discussed so far have been largely seasonal. However, secondary circulation also includes wind systems that migrate constantly, producing rapidly changing weather conditions throughout all seasons, especially in the middle latitudes. The migratory circulation systems are associated with air masses, fronts, cyclones, and anticyclones. These are covered in detail in the next unit.

Anticyclones

An anticyclone is an area of relatively high pressure with a clockwise flow in the Northern Hemisphere and counterclockwise flow in the Southern Hemisphere. The windflow in an anticyclone is slightly across the isobars and away from the center of the anticyclone.

Anticyclones are commonly called highs or high-pressure areas. The formation of an anticyclone or the intensification of an existing one is called *anticyclogenesis*. Anticyclogenesis refers to the development of anticyclonic circulation as well as the intensification of an existing anticyclonic flow. When a high-pressure center is increasing in pressure, the high is BUILDING or INTENSIFYING. Although a high can build without an increase in anticyclonic flow, it is rare. Normally, building and anticyclogenesis occur simultaneously. The weakening of anticyclonic circulation is *anticyclolysis*. When the pressure of a high is decreasing, we say the high is weakening. Anticyclolysis and weakening can occur separately, but usually occur together.

The vertical extent of pressure greatly depends on the air temperature. Since density increases with a decrease in temperature, pressure decreases more rapidly vertically in colder air than in warmer air. In a cold anticyclone, the vertical extent is shallow; while in a warm anticyclone, the vertical extent reaches high into the upper atmosphere due to the slow decrease in temperature with elevation.

Cyclones

A cyclone is a circular or nearly circular area of low pressure with a counterclockwise flow. The flow is slightly across the isobars toward the center in the Northern Hemisphere and clockwise in the Southern Hemisphere. It is commonly called a low or a depression. This use of the word cyclone should be distinguished from the colloquial use of the word as applied to the tornado or tropical cyclone.

The formation of a new cyclone or the intensification of the cyclonic flow in an existing one is called *cyclogenesis*. When the pressure in the low is falling, we say the low is deepening. Cyclogenesis and deepening can also occur separately, but usually occur at the same time. The decrease or eventual dissipation of a cyclonic flow is called *cyclolysis*. When the pressure in a low is rising, we say the low is filling. Cyclolysis and filling usually occur simultaneously. Cyclones in middle and high latitudes are referred to as extratropical cyclones. The term *tropical cyclone* refers to hurricanes and typhoons.

VERTICAL STRUCTURE OF SECONDARY CIRCULATIONS

To better understand the nature of the pressure centers of the secondary circulation, it is necessary to consider them from a three-dimensional standpoint. With the aid of surface and upper air charts, you will be able to see the three dimensions of these pressure systems as well as the circulation patterns of the secondary circulation as established at higher levels in the troposphere and lower stratosphere. In Chapter 2, the study of gas laws showed that volume is directly proportional to temperature. Stated another way, we might say that the thickness of a layer between two isobaric surfaces is directly proportional to the mean virtual temperature of the layer. Because the atmosphere is always moist to some degree, virtual temperature is used. Mean virtual temperature is defined as the average temperature at which dry air would have the same pressure and density as moist air.

Thus, lines representing thickness are also isotherms of mean virtual temperature. The higher the mean virtual temperature, the thicker the layer, or vice versa. The thickness between layers is expressed in geopotential meters. The shift in location, as well as the change of shape and intensity upward of atmospheric pressure systems, is dependent on the temperature distribution. An example of the effects of virtual temperature can be demonstrated by placing two columns of air next to each other. One air column is cold and the other air column is warm. The constant pressure surfaces in the cold column are closer than the ones in the warm column. Using the hypsometric equation can derive the thickness value between two pressure surfaces. Thickness may also be determined from tables, graphs, etc.

VERTICAL STRUCTURE OF HIGH PRESSURE SYSTEMS

The topographic features that indicate the circulation patterns at 500 millibars in the atmosphere correspond in general to those at lower and higher level. However, they may experience a shift in location as well as a change in intensity and shape. For example, a ridge aloft may reflect a closed high on a surface synoptic chart. In addition, upper air circulation patterns may take on a wavelike structure in contrast to the alternate closed lows, or closed high patterns at the surface level. The smoothing of the circulation pattern aloft is typical of atmospheric flow patterns.

Cold Core Highs

A cold core high is one in which the temperatures on a horizontal plane decrease toward the center. Because the temperature in the center of a cold core high is less than toward the outside of the system, it follows that the vertical spacing of isobars in the center of this system is closer together than on the outside. Although the pressure at the center of these systems on the surface may be high, the pressure decreases rapidly with height.

Because these highs are often quite shallow, it is common for an upper level low to exist above a cold core high. If the cold core high becomes subjected to warming from below and to subsidence from aloft, as it moves southward from its source and spreads out, it diminishes rapidly in intensity with time. Since these highs decrease in intensity with height, thickness is relatively low. In the vertical, cold core highs slope toward colder air aloft. Anticyclones found in Arctic air are always cold cored, while anticyclones in polar air may be warm or cold core.

Examples of cold core highs are the North American High, the Siberian High and the migratory highs that originate from these anticyclones.

Warm Core Highs

A warm core high is one in which the temperatures on a horizontal level increase toward the center. Because the temperatures in the center of a warm core high are higher than on the outside of the system, it follows that the vertical spacing of isobars in the center is farther apart than toward the outside of the high. For this reason, a warm core high increases in intensity with altitude and has an anticyclonic circulation at all levels.

From a vertical view, warm core highs slope toward warmer air aloft. A warm core high is accompanied by a high cold tropopause. Since the pressure surfaces are spaced far apart, the tropopause is reached only at great heights. The temperature continues to decrease with elevation and is cold by the time the tropopause is reached. The subtropical highs are good examples of this type of high. Therefore, anticyclones found in tropical air are alwayswarm core. Examples of warm core highs are the Azores or Bermuda High and the Pacific High.

VERTICAL STRUCTURE OF LOW-PRESSURE SYSTEMS

Low-pressure systems, like high-pressure systems, are generally a reflection of systems aloft. They, too, experience shifts in location and changes in intensity and shape with height. At times, a surface system may not be evident aloft and a well-developed system aloft may not reflect on a surface analysis.

Cold Core Lows

The cold core low contains the coldest air at its center throughout the troposphere; that is, going outward in any direction at any level in the troposphere, warmer air is encountered. The cold core low increases intensity with height. Relative minimums in thickness values, called cold pools, are found in such cyclones. The temperature distribution is almost symmetrical, and the axis of the low is nearly vertical. When they do slope vertically, they slope toward the coldest temperatures aloft. In the cold low, the lowest temperatures coincide with the lowest pressures.

The cold low has a more intense circulation aloft from 850 to 400 millibars than at the surface. Some cold lows show only slight evidence in the surface pressure field that an intense circulation exists aloft. The cyclonic circulation aloft is usually reflected on the surface in an abnormally low daily mean temperature often accompanied by instability and showery precipitation. A cold core low is accompanied by a low warm tropopause. Since the pressure surfaces are close together, the tropopause is reached at low altitudes where the temperature is relatively warm.

Good examples of cold core lows are the Aleutian and Icelandic lows. Occluded cyclones are generally cold core in their later stages, because polar or arctic air has closed in on them. At high latitudes the cold pools and their associated upper air lows show some tendency for location in the northern Pacific and Atlantic Oceans where, statistically, they contribute to the formation of the Aleutian and Icelandic lows.

Warm Core Lows

A warm core low decreases intensity with height and the temperature increases toward the center on a horizontal plane. The warm low is frequently stationary, such as the heat low over the southwestern United States in the summer; this is a result of strong heating in a region usually insulated from intrusions of cold air that tend to fill it or cause it to move. Thewarm lowis also found in its moving form as a stable wave moving along a frontal surface. There is no warm low aloft in the troposphere.

The tropical cyclone, however, is believed to be a warm low because its intensity diminishes with height. Because most warm lows are shallow, they have little slope. However, intense warm lows like the heat low over the southwest United States and hurricanes do slope toward warm air aloft. In general, the temperature field is quite asymmetrical around a warm core cyclone. Usually the southward moving air in the rear of the depression is not as warm as that moving northward in advance of it. A warm core low decreases intensity with height or completely disappears and are often replaced by anticyclones aloft. The heat lows of the southwestern United States, Asia, and Africa are good examples of warm core lows. Newly formed waves are generally warm core because of the wide-open warm sector.

DYNAMIC LOW

Systems that retain their closed circulations to appreciable altitudes and are migratory are called dynamic lows or highs. A dynamic low is a combination of a warm surface low and a cold upper low or trough, or a warm surface low in combination with a dynamic mechanism aloft for producing a cold upper low or trough. It has an axis that slopes toward the coldest tropospheric air.

In the final stage, after occlusion of the surface warm low is complete, the dynamic lowbecomes a cold lowwith the axis of the low becoming practically vertical.

DYNAMIC HIGH

The dynamic high is a combination of a surface cold high and an upper-level warm high or well-developed ridge, or a combination of a surface cold high with a dynamic mechanism aloft for producing high-level anticyclogenesis. Dynamic highs have axes that slope toward the warmest tropospheric air. In the final stages of warming the cold surface high, the dynamic high becomes a warm high with its axis practically vertical.

TERTIARY CIRCULATION

Tertiary circulations are localized circulations directly attributable to one of the following causes or a combination of them: local cooling, local heating, adjacent heating or cooling, and induction. Many regions have local weather phenomena caused by temperature differences between land and water surfaces or by local topographical features. These weather phenomena show up as circulations.

These tertiary circulations can result in dramatic local weather conditions and wind flows. The most common tertiary circulations are discussed in this lesson. However, there are numerous other circulations and related phenomena in existence around the world.

MONSOON WINDS

The term *monsoon* is of Arabic origin and means season. The monsoon wind is a seasonal wind that blows from continental interiors to the ocean in the winter; they blow in the opposite direction during the summer. The monsoon wind is most pronounced over India, although there are other regions with noticeable monsoon winds.

Monsoon winds are a result of unequal heating and cooling of land and water surfaces. During winter a massive area of cold high pressure develops over the extensive Asiatic continent. This high pressure is due primarily to cold arctic air and long-term radiation cooling. To the south, the warm equatorial waters exist and, in contrast, the area has relatively lower surface pressures. The combination of high pressure over Asia and low pressure over the Equatorial Belt sets up a pressure gradient directed from north to south. Because of the flow around the massive Siberian high, northeast winds begin to dominate the regions from India to the Philippines.

During the winter months, clear skies predominate over most of the region. This is caused by the mass motion of air from a high-pressure area over land to an area of lower pressure over the ocean. As the air leaves the high-pressure area over land, it is cold and dry. As it travels over land toward the ocean, there is no source of moisture to induce precipitation. The air is also traveling from a higher altitude to a lower altitude; consequently, this downslope motion causes the air to be warmed at the adiabatic lapse rate. This warming process has a still further clearing effect on the skies. During the summer the airflow over the region is completely reversed. The large interior of Asia is heated to the point where the continent is much warmer than the ocean areas to the south. This induces relatively low pressure over Asia and higher pressure over the equatorial region.

The weather associated with the summer monsoon winds is thunderstorms, almost constant heavy rain, rain showers, and gusty surface winds. This condition is caused by mass motion of air from the relatively high-pressure area over the ocean to a low-pressure area over land. When the air leaves the ocean, it is warm and moist. As the air travels over land toward the low-pressure area, it is also traveling from a lower altitude to a higher altitude. The air is lifted by a mechanical force and cooled to its condensation point by this upslope motion.

LAND AND SEA BREEZES

There is a diurnal contrast in the heating of local water and land areas similar to the seasonal variation of the monsoon. During the day, the land is warmer than the water area; at night the land area is cooler than the water area. A slight variation in pressure is caused by this temperature contrast. At night the wind blows from land to sea and is called a land breeze. During the day, the wind blows from water areas to land areas and is called a sea breeze.

The sea breeze usually begins during midmorning when the land areas become warmer than adjacent ocean waters. This temperature difference creates an area of slightly lower surface pressures over land compared to the now cooler waters. The result is a wind flow from water to land. The sea breeze starts with a shallow flow along the surface; however, as maximum heating occurs, the flow increases with height. The height varies from an average of 3,000 feet in moderately warm climates to 4,500 in tropical regions. The effects of the sea breeze can be felt as far as 30 miles both onshore and offshore. In extreme cases, the sea breeze is felt 100 miles inland depending upon terrain. By mid afternoon the sea breeze will reach its maximum speed and may be strong enough to be influenced by the Coriolis force, which causes it to flow at an angle to the shore. The sea breeze is most pronounced in late spring, summer, and early fall when maximum temperature differences occur between land and water surfaces. A decrease in temperature and an increase in humidity and wind speed mark the start of a sea breeze.

The sea breeze continues until the land area cools sufficiently to dissipate the weak low pressure. After sunset, the land cools at a faster rate than the adjacent waters and eventually produces a complete reversal of the winds. As the land continues to cool through the evening hours, a weak area of high pressure forms over the land. The water area, with its warmer temperatures, has slightly lower pressure and again a flow is established; however, the flow is now from land to Water.

The land breezes, when compared to the sea breezes, are less extensive and not as strong. This is because there is less temperature contrast at night between land and water surfaces as compared to the temperature contrast during daytime heating. Land breezes are at maximum development late at night, in late fall and early winter. In the tropical land regions, the land and sea breezes are repeated day after day with great regularity. In high latitudes the land and sea breezes are often masked by winds of synoptic features.

WINDS DUE TO LOCAL COOLING AND HEATING

In the next sections we discuss tertiary circulations due to local cooling and heating effects. Under normal circumstances, these winds attain only light to moderate wind speeds; however, winds often occur in and near mountain areas that have undergone dramatic changes in normal character. At times, mountain areas tend to funnel winds through valleys and mountain passes. This funneling effect produces extremely dangerous wind speeds.

FUNNEL EFFECT

Winds blowing against mountain barriers tend to flatten out and go around or over them. If a pass or a valley breaks the barrier, the air is forced through the break at considerable speed. When wind is forced through narrow valleys it is known as the funnel effect and is explained by Bernoulli's theorem. According to Bernoulli's theorem, pressures are least where velocities are greatest; likewise, pressures are greatest where velocities are least. This observation is true for both liquids and gases.

Bernoulli's theorem is frequently used to forecast tertiary winds in the mountainous western United States. The famous Santa Ana winds of southern California are a prime example. Winds associated with high pressure situated over Utah are funneled through the valley leading into the town of Santa Ana near the California coast.

Low pressure develops at the mouth of the valley and the end result is hot, dry, gusty and extremely dangerous winds. When the Santa Ana is strong enough, the effects are felt in virtually every valley located along the coast of southern California. Visibility is often restricted due to blowing sand. It is common to see campers, trailers, and trucks turned over by the force of these winds. When funneled winds reach this magnitude, they are called jet-effect winds, canyon winds, or mountain-gap winds.

Winds Due to Local Cooling and Heating

There are two types of tertiary circulations produced by local cooling—glacier winds and drainage winds.

GLACIER WINDS

Glacier winds, or fall winds occur in many varieties in all parts of the world where there are glaciers or elevated land masses that become covered by snow and ice during winter. During winter, the area of snow cover becomes most extensive. Weak pressure results in a maximum of radiation cooling. Consequently the air coming in contact with the cold snow cools. The cooling effect makes the overlying air more dense, therefore, heavier than the surrounding air. When set in motion, the cold dense air flows down the sides of the glacier or plateau. If it is funneled through a pass or valley, it may become very strong. This type of wind may form during the day or night due to radiation cooling. The glacier wind is most common during the winter when more snow and ice are present. When a changing pressure gradient moves a large cold air mass over the edge of a plateau, this action sets in motion the strongest, most persistent, and most extensive of the glacier or fall winds. When this happens, the fall velocity is added to the pressure gradient force causing the cold air to rush down to sea level along a front that may extend for hundreds of miles. This condition occurs in winter on a large scale along the edge of the Greenland icecap. In some places along the icecap, the wind attains a velocity in excess of 90 knots for days at a time and reaches more than 150 nautical miles out to sea.

Glacier winds are cold katabatic winds. Since all katabatic winds are heated adiabatically in their descent, they are predominantly dry.Occasionally, the glacier winds pick up moisture from falling precipitation when they underride warm air. Even with the adiabatic heating they undergo, all glacier or fall winds are essentially cold winds because of the extreme coldness of the air in their source region. Contrary to all other descending winds that are warm and dry, the glacier wind is cold and dry. It is colder, level for level, than the air mass it is displacing. In the Northern Hemisphere, the glacier winds descend frequently from the snow-covered plateaus and glaciers of Alaska, Canada, Greenland, and Norway.

DRAINAGE WINDS

Drainage winds are caused by the cooling air along the slopes of a mountain. Consequently, the air becomes heavy and flows downhill, producing the MOUNTAIN BREEZE. Drainage winds are katabatic winds and like glacier winds; a weak or nonexistent pressure gradient is required to start the downward flow. As the air near the top of a mountain cools through radiation or contact with colder surfaces, it becomes heavier than the surrounding air and gradually flows downward. Initially this flow is light and only a few feet thick. As cooling continues, the flow increases achieving speeds up to 15 knots at the base of the mountain and a depth of 200 feet or more. Winds in excess of 15 knots are rare and only occur when the mountain breeze is severely funneled. Drainage winds are cold and dry. Adiabatic heating does not sufficiently heat the descending air because of the relative coldness of the initial air and because the distance traveled by the air is normally short. Drainage winds have a very localized circulation. As the cold air enters the valley below, it displaces the warm air. Temperatures continue to fall. If the flow achieves speeds of 8 knots or more, mixing results between the warm valley air and the cold descending air that results in a slight temperature increase. Campers often prefer to make summer camps at the base of mountains to take advantage of the cooling effect of the mountain breeze.

Funnel Effects

VALLEY BREEZES

The valley breeze is the anabatic counterpart of the mountain breeze. When the sun heats the valley walls and mountain slopes during the morning hours, the air next to the ground is heated until it rises along the slopes. Rocky or sandy slopes devoid of vegetation are the most effective heating surfaces. If the slopes are steep, the ascending breeze tends to move up the valley walls. The expansion of the heated air next to the surface produces a slight local pressure gradient against the ground surface. As the heating becomes stronger, convective currents begin to rise vertically from the valleys.

The updrafts along the valley walls continue to be active, particularly at the head of the valley. The valley breeze usually reaches its maximum strength in the early afternoon. It is a stronger and deeper wind than the mountain breeze. It is difficult to isolate the valley breeze effect because of the prevailing gradient winds. Consequently, the valley breeze is much more likely to be superposed as a prevailing wind than is the mountain breeze, which by its very nature can develop only in the absence of any appreciable gradient wind. The valley breezes are generally restricted to slopes facing south or the more direct rays of the sun, and they are more pronounced in southern latitudes. They are diurnally strongest in the late afternoon and are seasonally strongest in summer.

THERMAL

Thermals are vertical convective currents that result from local heating. They stop short of the condensation level. Thermal convection is the usual result of strong heating of the lower atmosphere by the ground surface. A superadiabatic lapse rate immediately above the ground is necessary to the development of strong thermals. They form most readily over areas of bare rock or sand and in particular over sand dunes or bare rocky hills. In the presence of a moderate or fresh breeze, especially in a hilly terrain, it is impossible to distinguish between turbulent and thermal convection currents. Pure thermal convection normally occurs on clear summer days with very light prevailing wind. In the eastern United States, dry thermals are usually of only moderate intensity, seldom reaching an elevation in excess of 5,000 feet above the surface. The high moisture content of the air masses in this section in summer reduces the intensity of surface heating to some extent. This moisture content usually keeps the condensation level of the surface air near or even below a height of 5,000 feet above the ground. In the dry southwestern part of the country, where ground heating during clear summer days is extreme, dry thermal convection may extend to a height of 10,000 feet or more. Under these conditions, extremely turbulent air conditions can occur locally up to whatever heights the thermals extend, frequently without a cloud in the sky.

One variation of the dry thermal is seen in the dust or sand whirls, sometimes called dust devils. They are formed over heated surfaces when the winds are very light. Dust whirls are seldom more than two or three hundred feet high and they last only a few minutes at most. Over the desert on clear hot days as many as a dozen columns of whirling sand may be visible at once. The large desert sand whirls can become several hundred feet in diameter, extend to heights of 4,000 feet or higher, and in some cases last for an hour or more. They have been observed to rotate both anticyclonically and cyclonically, the same as tornadoes. An almost identical phenomenon is observed over water in the form of the waterspout.Waterspouts occur frequently in groups and form in relatively cool humid air over a warm water surface when the wind is light. The waterspout is visible due to the condensed water vapor, or cloud formation, within the vortex. The condensation is the result of dynamic cooling by expansion within the vortex. In this respect it differs from the sand whirl, which is always dry. Both the sand whirl and the waterspout represent simple thermal convection of an extreme type. They are not to be confused with the more violent tornado.

When dry thermal convection extends to an elevation where the dry thermals reach the condensation level, then cumulus convection takes the place of the dry convection. A cumulus cloud, whose base is at the condensation level of the rising air, tops each individual thermal current. Beneath every building cumulus cloud a vigorous rising current or updraft is observed. Thus the local thermal convection pattern becomes visible in the cumulus cloud pattern. The cumulus clouds form first over the hills where the strongest thermals develop. Under stable atmospheric conditions, little convective cloud development occurs. However, under unstable conditions these thermals may develop cumulonimbus clouds.

INDUCED OR DYNAMIC TERTIARY CIRCULATIONS

There are four types of induced or dynamic tertiary circulations. They are eddies, turbulence, large-scale vertical waves, and Foehn winds.

Eddies

An eddy is a circulation that develops when the wind flows over or adjacent to rough terrain, buildings, mountains or other obstructions. They generally form on the lee side of these obstructions. The size of the eddy is directly proportional to the size of the obstruction and speed of the wind. Eddies may have horizontal or vertical circulations that can be either cyclonic or anticyclonic. Horizontal eddies form in sheltered areas downwind of rough coastlines or mountain chains. An example of a horizontal eddy is the weak cyclonic circulation that develops in the channel off the coast of Santa Barbara, California. The winds frequently blow parallel to the northern California coastline during the winter fog and stratus season. The Santa Barbara channel often remains fog-free because the waters are protected from winds that transport the fog inland. However, when the winds are sufficiently strong, friction along the tough coastal range produces a weak cyclonic eddy over the channel. This cyclonic flow, though weak, is sufficient to advect fog into the region. Vertical eddies are generally found on the lee side of mountains, but with low wind speeds, stationary eddies or rotating pockets of air are produced and remain on both the windward and leeward sides of obstructions.

When wind speeds exceed about 20 knots, the flow may be broken up into irregular eddies that are carried along with a wind some distance downstream from the obstruction. These eddies may cause extreme and irregular variations in the wind and may disturb aircraft landing areas sufficiently to be a hazard.

A similar and much disturbed wind condition occurs when the wind blows over large obstructions such as mountain ridges. In such cases the wind blowing up the slope on the windward side is usually relatively smooth. On the leeward side the wind spills rapidly down the slope, setting up strong downdrafts and causing the air to be very turbulent.

These downdrafts can be very violent. Aircraft caught in these eddies could be forced to collide with the mountain peaks. This effect is also noticeable in the case of hills and bluffs, but is not as pronounced.

Turbulence

Turbulence is the irregular motion of the atmosphere caused by the air flowing over an uneven surface or by two currents of air flowing past each other in different directions or at different speeds. The main source of turbulence is the friction along the surface of Earth. This is called mechanical turbulence. Turbulence is also caused by irregular temperature distribution. The warmer air rises and the colder air descends, causing an irregular vertical motion of air; this is called thermal turbulence. Mechanical turbulence is intensified in unstable air and is weakened in stable air. These influences cause fluctuations in the wind with periods ranging from a few minutes to more than an hour. If these wind variations are strong, they are called wind squalls and are usually associated with convective clouds. They are an indication of approaching towering cumulus or cumulonimbus clouds.

Gustiness and turbulence are more or less synonymous. Gustiness is an irregularity in the wind speed that creates eddy currents disrupting the smooth airflow. Thus, the term *gust* is usually used in conjunction with sudden intermittent increases in the wind speed near the surface levels. Turbulence, on the other hand, is used with reference to levels above the surface. Gustiness can be measured; turbulence, however, unless encountered by aircraft equipped with a gust probe or an accelerometer, is usually estimated.

Large-Scale Vertical Waves

Mountain waves occur on the lee side of topographical barriers and occur when the wind-flow is strong, 25 knots or more, and the flow is roughly perpendicular to the mountain range. The structure of the barrier and the strength of the wind determines the amplitude and the type of the wave. The wind speed gradually increases, reaching a maximum near the summit. On passing the crest, the flowbreaks down into a much more complicated pattern with downdrafts predominating. Turbulence in varying degrees can be expected and is particularly severe in the lower levels; however, it can extend to the tropopause to a lesser degree. Proceeding downwind, some 5 to 10 miles from the summit, the airflow begins to ascend in a definite wave pattern. Additional waves, generally less intense than the primary wave, may form downwind. These are similar to the series of ripples that form downstream from a submerged rock in a swiftly flowing river. The distance between successive waves usually ranges from 2 to 10 miles, depending largely on the existing wind speed and the atmospheric stability. However, wavelengths up to 20 miles have been reported.

It is important to know how to identify a wave situation. Pilots must be briefed on this condition so they can avoid the wave hazards. Characteristic cloud forms peculiar to wave action provide the best means of visual identification. These clouds may occur singly or in layers at heights usually above 20,000 feet, and may be quite ragged when the airflow at that level is turbulent. The roll cloud forms at a lower level, generally near the height of the mountain ridge. The cap cloud, shown partially covering the mountain slope, must always be avoided in flight because of turbulence, concealed mountain peaks, and strong downdrafts on the lee side. The lenticular, like the roll clouds and cap clouds, are stationary, constantly forming on the windward side and dissipating on the lee side of the wave. The actual cloud forms can be a guide to the degree of turbulence. Smooth clouds generally show smoother airflow in or near them with light turbulence. Clouds appearing ragged or irregular indicate more turbulence. While clouds are generally present to forewarn the presence of wave activity, it is possible for wave action to take place when the air is too dry to form clouds. This makes the problem of identifying and forecasting more difficult.

Foehn Winds

When air flows downhill from a high elevation, its temperature is raised by adiabatic compression. Foehn winds are katabatic winds caused by adiabatic heating of air as it descends on the lee sides of mountains. Foehn winds occur frequently in our western mountain states and in Europe in the late fall and winter. In Montana and Wyoming, the Chinook is a well-known phenomenon; in southern California, the Santa Ana is known particularly for its high-speed winds that easily exceed 50 knots. For the purpose of illustrating a Foehn wind, the Santa Ana is used.

The condition producing the Foehn wind is a high-pressure area with a strong pressure gradient situated near Salt Lake City, Utah. This gradient directs the wind flow into a valley leading to the town of Santa Ana near the coast of California. As the wind enters the valley, its flow is sharply restricted by the funneling effect of the mountainsides. This restriction causes the wind speed to increase, bringing about a drop in pressure in and near the valley. The Bernoulli effect causes this pressure drop in and near a valley. Generally speaking, when the Santa Ana blows through the Santa Ana Canyon, a similar wind simultaneously affects the entire southern California area. Thus, when meteorological conditions are favorable, this dry northeast wind blows through the many passes and canyons, over all the mountainous area, including the highest peaks, and quite often at exposed places along the entire coast from Santa Barbara to San Diego. Therefore, the term Santa Ana refers to the general condition of a dry northeast wind over southern California.

In the Rocky Mountain states, the onset of Foehn winds have accounted for temperature rises of 50°F or more in only a few minutes. In southern California, the temperature, though less dramatically, also rises rapidly and is accompanied by a rapid decrease in humidity and a strong shift and increase in wind speeds. Although these winds may on occasion reach destructive velocities, one beneficial aspect is that these winds quickly disperse the severe air pollutants that plague the Los Angeles Basin.

PLANETARY AND LOCAL WINDS

PLANETARY WIND

The hydrodynamic wind within the upper portion of a planet's atmosphere allows light chemical elements such as hydrogen to move up to the exobase, the lower limit of the exosphere, where the gases can then reach escape velocity, entering outer space without impacting other particles of gas. This type of gas loss from a planet into space is known as planetary wind. Such a process over geologic time causes water-rich planets such as the Earth to evolve into planets such as Venus over billions of years. Planets with hot lower atmospheres could result in humid upper atmospheres that accelerate the loss of hydrogen.

LOCAL WINDS

Local winds are winds that blow over short distances caused by unequal heating of the earths surface in a small area. Local winds blow over a much smaller area than global winds and have a much shorter time span. Hot windsoriginate in vast anticyclones over hot deserts and include the Santa Ana, the Brickfielder, the Sirocco, the Haboob, the Khamsin, and the Harmattan.

MONSOONS AND JET STREAM

MONSOON

Monsoon is traditionally defined as a seasonal reversing wind accompanied by corresponding changes in precipitation, but is now used to describe seasonal changes in atmospheric circulation and precipitation associated with the asymmetric heating of land and sea. Usually, the term monsoon is used to refer to the rainy phase of a seasonally-changing pattern, although technically there is also a dry phase.

The major monsoon systems of the world consist of the West African and Asia-Australian monsoons. The inclusion of the North and South American monsoons with incomplete wind reversal may be debated.

The term was first used in English in British India and neighbouring countries to refer to the big seasonal winds blowing from the Bay of Bengal and Arabian Sea in the southwest bringing heavy rainfall to the area.

Process

Monsoons may be considered as large-scale sea breezes, due to seasonal heating and the resulting development of a thermal low over a continental landmass. They are caused by the larger amplitude of the seasonal cycle of land temperature compared to that of nearby oceans. This differential warming happens because heat in the ocean is mixed vertically through a "mixed layer" that may be fifty metres deep, through the action of wind and buoyancy-generated turbulence, whereas the land surface conducts heat slowly, with the seasonal signal penetrating perhaps a metre or so. Additionally, the specific heat capacity of liquid water is significantly higher than that of most materials that make up land. Together, these factors mean that the heat capacity of the layer participating in the seasonal cycle is much larger over the oceans than over land, with the consequence that the air over the land warms faster and reaches a higher temperature than the air over the ocean. The hot air over the land tends to rise, creating an area of low pressure. This creates a steady wind blowing toward the land, bringing the moist near-surface air over the oceans with it. Similar rainfall is caused by the moist ocean air being lifted upwards by mountains, surface heating, convergence at the surface, divergence aloft, or from storm-produced outflows at the surface. However the lifting occurs, the air cools due to expansion in lower pressure, which in turn produces condensation.

In winter, the land cools off quickly, but the ocean retains heat longer. The cold air over the land creates a high pressure area which produces a breeze from land to ocean. Monsoons are similar to sea and land breezes, a term usually referring to the localized, diurnal cycle of circulation near coastlines, but they are much larger in scale, stronger and seasonal.

Most summer monsoons have a dominant westerly component and a strong tendency to ascend and produce copious amounts of rain. The intensity and duration, however, are not uniform from year to year. Winter monsoons, by contrast, have a dominant easterly component and a strong tendency to diverge, subside and cause drought. Even more broadly, it is now understood that in the geological past, monsoon systems likely accompanied the formation of supercontinents such as Pangaea, with their extreme continental climates.

JET STREAM

Jet streams are fast flowing, narrow air currents found in the atmospheres of some planets, including Earth. The main jet streams are located near the tropopause, the transition between the troposphere and the stratosphere. The major jet streams on Earth are westerly winds. Their paths typically have a meandering shape; jet streams may start, stop, split into two or more parts, combine into one stream, or flow in various directions including the opposite direction of most of the jet.

The strongest jet streams are the polar jets, at around 7–12 km above sea level, and the higher and somewhat weaker subtropical jets at around 10–16 km. The northern hemisphere and the southern hemisphere each have both a polar jet and a subtropical jet. The northern hemisphere polar jet flows over the middle to northern latitudes of North America, Europe, and Asia and their intervening oceans, while the southern hemisphere polar jet mostly circles Antarctica all year round.

Jet streams are caused by a combination of a planet's rotation on its axis and atmospheric heating. The Coriolis effect describes how a planet's surface and atmosphere rotate fastest at the planet's equator while virtually not rotating at all at the poles. Jet streams form near boundaries of adjacent air masses with significant differences in temperature, such as the polar region and the warmer air towards the equator. Meteorologists use the location of some of the jet streams as an aid in weather forecasting. The main commercial relevance of the jet streams is in air travel, as flight time can be dramatically affected by either flying with the flow or against the flow of a jet stream. Clear-air turbulence, a potential hazard to aircraft passenger safety, often is found in a jet stream's vicinity. One future benefit of jet streams could be to power airborne wind turbines.

Other jets also exist. During the northern hemisphere summer, easterly jets can form in tropical regions, typically in a region where dry air encounters more humid air at high altitudes. Low-level jets also are typical of various regions such as the central United States.

Discovery

After the 1883 eruption of the Krakatoa volcano, weather watchers tracked and mapped the effects on the sky over several years. They labeled the phenomenon the "equatorial smoke stream." In the 1920s, a Japanese meteorologist, Oishi Wasaburo, detected the jet stream from a site near Mount Fuji. He tracked pilot balloons, also known as pibals, as they rose into the atmosphere. Oishi's work largely went unnoticed outside Japan. American pilot Wiley Post, the first man to fly around the world solo in 1933, is often given some credit for discovery of jet streams.

Post invented a pressurized suit that let him fly above 6,200 metres. In the year before his death, Post made several attempts at a high-altitude transcontinental flight, and noticed that at times his ground speed greatly exceeded his air speed. German meteorologist H. Seilkopf is credited with coining a special term, *Strahlströmung* for the phenomenon in 1939. Many sources credit real understanding of the nature of jet streams to regular and repeated flight-path traversals during World War II. Flyers consistently noticed westerly tailwinds in excess of 100 mph in flights, for example, from the US to the UK.

Description

Polar jet streams are typically located near the 250 hPa pressure level, or 7 kilometres to 12 kilometres above sea level, while the weaker subtropical jet streams are much higher, between 10 kilometres and 16 kilometres above sea level. In each hemisphere, both upper-level jet streams form near breaks in the tropopause, which is at a higher altitude near the equator than it is over the poles, with large changes in its height occurring near the location of the jet stream.

The northern hemisphere polar jet stream is most commonly found between latitudes 30°N and 60°N, while the northern subtropical jet stream located close to latitude 30°N. The upper level jet stream is said to "follow the sun" as it moves northward during the warm season, or late spring and summer, and southward during the cold season, or autumn and winter. The width of a jet stream is typically a few hundred miles and its vertical thickness often less than three miles.

Jet streams are typically continuous over long distances, but discontinuities are common. The path of the jet typically has a meandering shape, and these meanders themselves propagate east, at lower speeds than that of the actual wind within the flow. Each large meander, or wave, within the jet stream is known as a Rossby wave. Rossby waves are caused by changes in the Coriolis effect with latitude, and propagate westward with respect to the flow in which they are embedded, which slows down the eastward migration of upper level troughs and ridges across the globe when compared to their embedded shortwave troughs. Shortwave troughs are smaller packets of upper level energy, on the scale of 1,000 kilometres to 4,000 kilometres long, which move through the flow pattern around large scale, or longwave, ridges and troughs within Rossby waves. Jet streams can split into two due to the formation of an upper-level closed low, which diverts a portion of the jet stream under its base, while the remainder of the jet moves by to its north.

The wind speeds vary according to the temperature gradient, exceeding 92 kilometres per hour, although speeds of over 398 kilometres per hour have been measured. Meteorologists now understand that the path of jet streams steers cyclonic storm systems at lower levels in the atmosphere, and so knowledge of their course has become an important part of weather forecasting. For example, in 2007, Britain experienced severe flooding as a result of the polar jet staying south for the summer. The polar and subtropical jets merge at some locations and times, while at other times they are well separated.

Cause

In general, winds are strongest under the tropopause. If two air masses of different temperatures or densities meet, the resulting pressure difference caused by the density difference is highest within the transition zone. The wind does not flow directly from the hot to the cold area, but is deflected by the Coriolis effect and flows along the boundary of the two air masses. All these facts are consequences of the thermal wind relation.

The balance of forces on an atmospheric parcel in the vertical direction is primarily between the pressure gradient and the force of gravity, a balance referred to as hydrostatic. In the horizontal, the dominant balance outside of the tropics is between the Coriolis effect and the pressure gradient, a balance referred to as geostrophic. Given both hydrostatic and geostrophic balance, one can derive the thermal wind relation: the vertical derivative of the horizontal wind is proportional to the horizontal temperature gradient. The sense of the relation is such that temperatures decreasing polewards implies that winds develop a larger eastward component as one moves upwards. Therefore, the strong eastward moving jet streams are in part a simple consequence of the fact that the equator is warmer than the north and south poles.

The thermal wind relation does not immediately provide an explanation for why the winds are organized in tight jets, rather than distributed more broadly over the hemisphere. There are two factors that contribute to this sharpness of the jets. The first factor is the tendency for developing cyclonic disturbances in midlatitudes to form fronts — sharp localized gradients in temperature. The polar front jet stream can be thought of as the result of this frontogenesis process in midlatitudes, as the storms concentrate the north-south temperature contrast into relatively narrow regions.

Polar jet

The polar jet stream can be thought of as the result of this frontogenesis process in midlatitudes.

Subtropical jet

A second factor which contributes to jet sharpness is more appropriate for the subtropical jet. The subtropical jet forms at the poleward limit of the tropical Hadley cell and to first order this circulation is symmetric with respect to longitude. Tropical air rises to the tropopause, mainly because of thunderstorm systems in the Intertropical Convergence Zone, and moves poleward before sinking; this is the Hadley circulation. As it does so it tends to conserve angular momentum, since friction is slight above the ground. In the northern hemisphere motions are deflected to the right by the Coriolis force, which for poleward moving air implies an increased eastward component of the winds. Around 30 degrees from the equator the jet wind speeds have become strong enough that were the jet to extend further polewards the increased windspeed would be unstable; thus the jet is limited.

Other planets

Jupiter's atmosphere has multiple jet streams, forming the familiar banded color structure, caused by internal heating. The factors that control the number of jet streams in a planetary atmosphere is an active area of research in dynamical meteorology. In models, as one increases the planetary radius, holding all other parameters fixed, the number of jet streams increases.

Uses

On Earth, the northern polar jet stream is the most important one for aviation and weather forecasting, as it is much stronger and at a much lower altitude than the subtropical jet streams and also covers many countries in the Northern Hemisphere, while the southern polar jet stream mostly circles

Antarctica and sometimes the southern tip of South America. The term *jet stream* in these contexts thus usually implies the northern polar jet stream.

Aviation

The location of the jet stream is extremely important for aviation. Commercial use of the jet stream began on November 18, 1952, when Pan Am flew from Tokyo to Honolulu at an altitude of 7,600 metres. It cut the trip time by over one-third, from 18 to 11.5 hours. Not only does it cut time off the flight, it also nets fuel savings for the airline industry. Within North America, the time needed to fly east across the continent can be decreased by about 30 minutes if an airplane can fly with the jet stream, or increased by more than that amount if it must fly west against it.

Associated with jet streams is a phenomenon known as clear air turbulence caused by vertical and horizontal windshear connected to the jet streams. The CAT is strongest on the cold air side of the jet, next to and just underneath the axis of the jet. Clear air turbulence can cause aircraft to plunge and so present a passenger safety hazard that has caused fatal accidents, such as the death of one passenger on United Airlines Flight 826.

Future power generation

Scientists are investigating ways to harness the wind energy within the jet stream. According to one estimate, of the potential wind energy in the jet stream, only 1 percent would be needed to meet the world's current energy needs. The required technology would reportedly take 10–20 years to develop.

Unpowered aerial attack

Towards the end of World War II the Japanese fire balloon was designed as a cheap weapon intended to make use of the jet stream over the Pacific Ocean to reach the west coast of Canada and the United States. They were relatively ineffective as weapons but were used in one of the few attacks on North America during World War II, causing six deaths and a small amount of damage.

Changes due to climate cycles

Effects of ENSO

The El Niño-Southern Oscillation influences the average location of upper-level jet streams, and leads to cyclical variations in precipitation and temperature across North America, as well as affecting tropical cyclone development across the eastern Pacific and Atlantic basins. Combined with the Pacific Decadal Oscillation, ENSO can also impact cold season rainfall in Europe. Changes in ENSO also change the location of the jet stream over South America, which partially affects precipitation distribution over the continent.

El Niño

During El Niño events, increased precipitation is expected in California due to a more southerly, zonal, storm track. During the El Niño portion of ENSO, increased precipitation falls along the Gulf coast and Southeast due to a stronger than normal, and more southerly, polar jet stream. Snowfall is greater than average across the southern Rockies and Sierra Nevada mountain range, and is well-below normal across the Upper Midwest and Great Lakes states.

The northern tier of the lower 48 exhibits above normal temperatures during the fall and winter, while the Gulf coast experiences below normal temperatures during the winter season. The subtropical jet stream across the deep tropics of the Northern Hemisphere is enhanced due to increased convection in the equatorial Pacific, which decreases tropical cyclogenesis within the Atlantic tropics below what is normal, and increases tropical cyclone activity across the eastern Pacific. In the Southern Hemisphere, the subtropical jet stream is displaced equatorward, or north, of its normal position, which diverts frontal systems and thunderstorm complexes from reaching central portions of the continent.

La Niña

Across North America during La Niña, increased precipitation is diverted into the Pacific Northwest due to a more northerly storm track and jet stream. The storm track shifts far enough northward to bring wetter than normal conditions to the Midwestern states, as well as hot and dry summers. Snowfall is above normal across the Pacific Northwest and western Great Lakes. Across the North Atlantic, the jet stream is stronger than normal, which directs stronger systems with increased precipitation towards Europe.

The Dust Bowl

Evidence suggests the jet stream was at least partially responsible for the widespread drought conditions during the 1930s Dust Bowl in the Midwest United States. Normally, the jet stream flows east over the Gulf of Mexico and turns northward pulling up moisture and dumping rain onto the Great Plains. During the Dust Bowl, the jet stream weakened and changed course traveling farther south than normal. This starved the Great Plains and other areas of the Midwest of precious rain creating dusty conditions.

Longer-term climatic changes

During 2007, 2008, 2009 the Jet Stream has been at an abnormally low latitude across the UK, lying closer to the English Channel, around 50°N rather than its more usual north of Scotland latitude of around 60°N. However, between 1979 and 2001, it has been found that the position of the jet stream has been moving northward at a rate of 2.01 kilometres per year across the Northern Hemisphere. Across North America, this type of change could lead to drier conditions across the southern tier of the United States and more frequent and more intense tropical cyclones in the tropics. A similar slow poleward drift was found when studying the Southern Hemisphere jet stream over the same time frame.

Polar night jet

The polar-night jet stream forms only during the winter months of the year in their respective hemispheres at around 60° latitude, but at a greater height than the polar jet, of about 80,000 feet. During these dark months the air high over the poles becomes much colder than the air over the Equator. This difference in temperature gives rise to extreme air-pressure differences in the stratosphere which, when combined with the Coriolis effect, create the polar night jets which race eastward at altitudes of about 30 miles.

Inside the polar night jet is the polar vortex. The warmer air can only move along the edge of the polar vortex, but not enter it. Within the vortex, the cold polar air becomes cooler and cooler with neither warmer air from lower latitudes nor energy from the sun during the polar night.

Low level jets

Barrier jet

A barrier jet in the low levels forms just upstream of mountain chains, with the mountains forcing the jet to be oriented parallel to the mountains. The mountain barrier increases the strength of the low level wind by 45 percent. In the North American Great Plains a southerly low-level jet helps fuel overnight thunderstorm activity during the warm season, normally in the form of mesoscale convective systems which form during the overnight hours. A similar phenomenon develops across Australia, which pulls moisture poleward from the Coral Sea towards cut-off lows which form mainly across southwestern portions of the continent.

Valley exit jet

A valley exit jet is a strong, down-valley, elevated air current that emerges above the intersection of the valley and its adjacent plain. These winds frequently reach a maximum of 20 m/s at a height of 40-200m above the ground. Surface winds below the jet may sway vegetation but are significantly weaker.

They are likely to be found in valley regions that exhibit diurnal mountain wind systems, such as those of the dry mountain ranges of the US. Deep valleys that terminate abruptly at a plain are more impacted by these factors than are those that gradually become shallower as downvalley distance increases.

Africa

The mid-level African easterly jet occurs during the Northern Hemisphere summer between 10°N and 20°N above West Africa, and the nocturnal poleward low-level jet occurs in the Great Plains of east and south Africa. The low-level easterly African jet stream is considered to play a crucial role in the southwest monsoon of Africa, and helps form the tropical waves which march across the tropical Atlantic and eastern Pacific oceans during the warm season. The formation of the thermal low over northern Africa leads to a low-level westerly jet stream from June into October.

AIR MASSES AND FRONTS

AIR MASS

In meteorology, an air mass is a volume of air defined by its temperature and water vapor content. Air masses cover many hundreds or thousands of square miles, and adopt the characteristics of the surface below them. They are classified according to latitude and their continental or maritime source regions. Colder air masses are termed polar or arctic, while warmer air masses are deemed tropical.

Continental and superior air masses are dry while maritime and monsoon air masses are moist. Weather fronts separate air masses with different density characteristics. Once an air mass moves away from its source region, underlying vegetation and water bodies can quickly modify its character. Classification schemes tackle an air mass' characteristics, and well as modification...

Classification and notation

The Bergeron classification is the most widely accepted form of air mass classification, though others have produced more refined versions of this scheme over different regions of the globe. Air mass classification involves three letters. The first letter describes its moisture properties, with c used for continental air masses and m for maritime air masses. The second letter describes the thermal characteristic of its source region: T for Tropical, P for Polar, A for arctic or Antarctic, M for monsoon, E for Equatorial, and S for superior air. The third letter is used to designate the stability of the atmosphere. If the air mass is colder than the ground below it, it is labeled k. If the air mass is warmer than the ground below it, it is labeled w.

Air masses of oceanic origin are denoted with a lower-case "m" ("maritime"), while air masses of continental origin are denoted with a lower-case "c" ("continental"). Air masses are also denoted as either Arctic, polar, tropical, or equatorial. These two sets of attributes are used in combinations depending on the air mass being described. For instance, an air mass originating over the desert southwest of the United States in summer may be designated "cT". An air mass originating over northern Siberia in winter may be indicated as "cA". An upper case "S" was occasionally used to denote something called a "superior" air mass. This was regarded as an adiabatically drying and warming air mass descending from aloft. In South Asia, an upper case "M" has been occasionally used to denote an air mass within the summer monsoon regime in that region.

The stability of an air mass may be shown using a third letter, either "k" or "w". An example of this might be a polar air mass blowing over the Gulf Stream, denoted as "cPk". Occasionally, one may also encounter the use of an apostrophe or "degree tick" denoting that a given air mass having the same notation as another it is replacing is colder than the replaced air mass. For example, a series of fronts over the Pacific might show an air mass denoted mPk followed by another denoted mPk'.

Another convention utilizing these symbols is the indication of modification or transformation of one type to another. For instance, an Arctic air mass blowing out over the Gulf of Alaska may be shown as "cA-mPk". Yet another convention indicates the layering of air masses in certain situations. For instance, the overrunning of a polar air mass by an air mass from the Gulf of Mexico over the Central United States might be shown with the notation "mT/cP".

Characteristics

Arctic, Antarctic, and polar air masses are cold. The qualities of arctic air are developed over ice and snow-covered ground. Arctic air is deeply cold, colder than polar air masses. Arctic air can be shallow in the summer, and rapidly modify as it moves equatorward. Polar air masses develop over higher latitudes over the land or ocean, are very stable, and generally shallower than arctic air. Polar air over the ocean loses its stability as it gains moisture over warmer ocean waters. Tropical and equatorial air masses are hot as they develop over lower latitudes. Those that develop over land are drier and hotter than those that develop over oceans, and travel northward on the western periphery of the subtropical ridge. Maritime tropical air masses are sometimes referred to as trade air masses.

Monsoon air masses are moist and unstable. Superior air masses are dry, and rarely reach the ground. It normally resides over maritime tropical air masses, forming a warmer and drier layer over the more moderate moist air mass below, forming what is known as a trade wind inversion over the maritime tropical air mass. Continental Polar air masses are air masses that are cold and dry due to their continental source region. Continental polar air masses that affect North America form over interior Canada. A Continental Tropical Air Mass is a type of tropical air produced over subtropical arid regions; it is hot and very dry.

Movement and fronts

A weather front is a boundary separating two masses of air of different densities, and is the principal cause of meteorological phenomena. In surface weather analyses, fronts are depicted using various colored lines and symbols, depending on the type of front. The air masses separated by a front usually differ in temperature and humidity. Cold fronts may feature narrow bands of thunderstorms and severe weather, and may on occasion be preceded by squall lines or dry lines. Warm fronts are usually preceded by stratiform precipitation and fog. The weather usually clears quickly after a front's passage. Some fronts produce no precipitation and little cloudiness, although there is invariably a wind shift.

Cold fronts and occluded fronts generally move from west to east, while warm fronts move poleward. Because of the greater density of air in their wake, cold fronts and cold occlusions move faster than warm fronts and warm occlusions. Mountains and warm bodies of water can slow the movement of fronts. When a front becomes stationary, and the density contrast across the frontal boundary vanishes, the front can degenerate into a line which separates regions of differing wind velocity, known as a shearline. This is most common over the open ocean.

Modification

Air masses can be modified in a variety of ways. Surface flux from underlying vegetation, such as forest, acts to moisten the overlying air mass. Heat from underlying warmer waters can significantly modify an air mass over distances as short as 35 kilometres to 40 kilometres. For example, southwest of extratropical cyclones curved cyclonic flow bringing cold air across the relatively warm water bodies can lead to narrow lake-effect snow bands.

Those bands bring strong localized precipitation since large water bodies such as lakes efficiently store heat that results in significant temperature differences between the water surface and the air above. Because of this temperature difference, warmth and moisture are transported upward, condensing into vertically oriented clouds which produce snow showers. The temperature decrease with height and cloud depth are directly affected by both the water temperature and the large-scale environment. The stronger the temperature decrease with height, the deeper the clouds get, and the greater the precipitation rate becomes.

TEMPERATURE AND TROPICAL CYCLONES

A tropical cyclone is composed of a system of thunderstorms that shows a cyclonic rotation around a central core or eye. A *tropical cyclone* is a generic term for a storm with an organized system of thunderstorms that are not based on a frontal system. Each individual tropical cyclone differs, but several characteristics are common to most all tropical cyclones including a central low-pressure zone and high wind speeds of at least 34 knots. At this point, the storms are given a pre-determined storm name. Most storms are accompanied by a lot of rain and storm surges near the shore. Often, once the storms make landfall, the tropical cyclone can cause tornadoes.

A tropical cyclone needs warm ocean temperatures in order to form. Temperatures in the ocean need to be at least 82 degrees Fahrenheit in order to form. Heat is drawn up from the oceans creating what is popularly called a 'heat engine'. Tall convective towers of clouds are formed within the storm as warm ocean water evaporates. As the air rises higher it cools and condenses releasing latent heat which causes even more clouds to form and feed the storm.

ROTATION AND FORWARD SPEED

The rotation of tropical cyclones in the Northern Hemisphere is counter-clockwise due to the Coriolis Effect. The opposite is true in the Southern Hemisphere.

The forward speed of a tropical cyclone can be a factor in determining the amount of damage the storm will cause. If a storm remains over one area for a long period of time, torrential rains, high winds, and flooding can severely impact an area. The average forward speed of a tropical cyclone is dependent on the latitude where the storm is currently. Generally, at less than 30 degrees of latitude, the storms will move at about 20 mph on average. The closer the storm is located the the equator, the slower the movement. Some storms will even stall out over an area for an extended period of time. After about 35 degrees North latitude, the storms start to pick up speed.

A good example of the fast formation of tropical cyclones comes when several storms stack up in the ocean back-to-back. Such an example occurred in 2009 with the formation of Ana, Bill, and Claudette as seen in this satellite image. The storms were very close to one another. Storms can also become entangle with one another in a process known as the Fujiwhara Effect where tropical cyclones can interact with each other.

NAMING TROPICAL CYCLONES

In the Atlantic Ocean and Eastern Pacific, tropical cyclones are known as hurricanes. In the Western Pacific Ocean, tropical cyclones are known as typhoons. In the Indian Ocean, a tropical cyclone is called a cyclone. These names are described in the article – Is it a typhoon, a cyclone, or a hurricane?

Specific storm names in each of the ocean basins vary based on conventional naming practices. For instance, in the Atlantic Ocean, storms are given names based on an alphabetical pre-determined list of Atlantic hurricane names. Severe hurricanes names are often retired.

TYPES AND DISTRIBUTION OF PRECIPITATION

The global distribution of precipitation is influenced by the general circulation of the atmosphere, proximity to large bodies of water, and topography. Precipitation is most abundant where air rises, and least abundant where it sinks. It also tends to be greater near oceans and lakes, and in higher elevations.

GENERAL CIRCULATION

The Earth's atmosphere is known to have regions characterized by largescale rising air, and other regions with descending air; these vary by latitude and by season. Rising air is found primarily near the equator and in the midlatitudes, so these tend to be wet areas. Descending air dominates in the subtropics and the poles. The global distribution of precipitation shows that the wettest areas on Earth are in the "rising air" zones, while the driest areas are in the "descending air" belts. As the Earth revolves around the Sun during the year, the orientation of its axis relative to the Sun changes. This causes the apparent position of the Sun relative to the Earth to change, and creates distinct seasons. Between March and September, the axis of Earth is tilted toward the Sun, and hence the Sun shines more directly over the Northern Hemisphere, resulting in more sunlight, more heat, and the warmer temperatures of Northern summer. In the other 6 months, the Earth's axis is tilted away from the Sun, and the Sun shines more directly over the Southern Hemisphere, bringing summer to countries south of the Equator.

The "rising" and "sinking" zones move northward and southward with the Sun's path. Thus, the wet area near the Equator moves northward into the Northern Hemisphere in its summer, and southward into the Southern Hemisphere during its summer. Similarly, the dry zones and wet zone at higher latitudes shift northward and southward throughout the year.

The result of these shifting zones are latitude bands with distinctive precipitation characteristics:

- 0–5° latitude: wet throught the year (rising zone)
- 5–20° latitude: wet summer (rising zone), dry winter (sinking zone)
- 20–30° latitude: dry all year (sinking zone)
- 30–50° latitude: wet winter (rising zone), dry summer (sinking zone)
- 50–60° latitude: wet all year (rising zone)
- 60–70° latitude: wet summer (rising zone), dry winter (sinking zone)
- 70–90° latitude: dry all year (sinking zone)

If the Earth had no mountains, and oceans were homogeneous with respect to their heat content, the climate would occur in latitude bands like those listed above. However, mountains indeed exist, and they exert a strong influence on precipitation, as do warm and cold ocean currents.

TOPOGRAPHY

When moving encounters a hill or mountain, it is forced to rise. Because rising air cools and condenses, precipitation is heaviest on the upwind side of a mountain, where the air is rising. This process is known as orographic lifting. On the downwind side, air descends, warms, and becomes drier. In some parts of the world winds are steady throughout the year, and if these steady, moist winds encounter a mountain range, precipitation will occur frequently.

One example of a mountainous area receiving frequent rain is Mt. Waialeale on Kauai, Hawaii. It is a sharp peak directly in the path of steady trade winds which blow from the northeast most of the time. On the upwind side of Waialeale, the air rises and condenses, resulting in almost constant clouds—nearly every day experiences rain. Waialeale is among the wettest places on Earth: over a 10-year period, annual rainfall averages more than 1,143 centimeters. This is more than 8 times greater than rainfall experienced in many parts of the midwestern United States.

In contrast, on the downwind side of Mt. Waialeale, the air descends, warms, and dries, in an area known as a "rain shadow." The result is a semiarid area with less than 51 centimeters of rain a year. This tremendous difference in precipitation—from 1,143 to 51 centimeters annually—occurs in a span of only 32 kilometers.

Not every location receives such steady winds as Waialeale. But most mountainous areas have a wet side and a dry side, depending on the typical wind direction. Mountains can partially or completely override the rising. The formation of clouds and precipitation is a function of latitude, topography, and the presence of warm or cold oceans or large lakes, among other factors.

This cumulus cloud is building into a cumulonimbus, which can bring heavy thunderstorms and even tornadoes. and descending zones described above. Mountains can bring precipitation to normally dry areas, and downwind slopes of mountains can be dry, even in normally wet areas like the tropics.

PRESENCE OF OCEANS OR LAKES

According to the water cycle concept, water originates in the ocean, evaporates, rises into the air, condenses, falls as precipitation, and returns to the ocean. This concept is generally true. However, the temperature of the water exerts a strong influence on how well the cycle operates.

The warmer the surface of the ocean, the faster it evaporates, all other things being equal. Warm ocean waters evaporate tremendous amounts of water into the atmosphere, causing the dew point to be very high. Near warm waterbodies, the air usually has enough moisture to produce abundant precipitation—all it takes is for the air to rise, either by heating, or by encountering a mountain or other elevated terrain. On the other hand, cold ocean currents are relatively passive when it comes to evaporating water, so the dew points in such areas are lower than near warm water.

As an example, consider two U.S. cities, both on the ocean at about the same latitude. Santa Monica, California, on the cool Pacific, is in the "wet winter, dry summer" zone between 30° and 50° North latitude, and receives only 36 centimeters of precipitation per year due to relatively low evaporation from the Pacific. Charleston, South Carolina, on the other hand, is on the very warm Atlantic. So much moisture is available to Charleston that it receives more rain in 2 months than Santa Monica does all year, even though Charleston should also be a "wet winter, dry summer" location. For the year, nearly 127 centimeters are recorded in Charleston.

Moreover, the high dew points yield very high humidity at Charleston and other eastern U.S. cities, making them more uncomfortable during summer. Meanwhile, in the "low dew point" city of Santa Monica, the residents are cooled by ocean breezes and enjoy steady sunshine during the typically very dry summers.

KÖPPEN AND THORNTHWAITE'S CLASSIFICATION OF WORLD CLIMATE

KÖPPEN CLIMATE CLASSIFICATION

The Köppen climate classification is one of the most widely used climate classification systems. It was first published by Russian German climatologist Wladimir Köppen in 1884, with several later modifications by Köppen himself notably in 1918 and 1936.

Later, German climatologist Rudolf Geiger collaborated with Köppen on changes to the classification system, which is thus sometimes referred to as the Köppen–Geiger climate classification system. The system is based on the concept that native vegetation is the best expression of climate. Thus, climate zone boundaries have been selected with vegetation distribution in mind. It combines average annual and monthly temperatures and precipitation, and the seasonality of precipitation.

Scheme

Köppen climate classification scheme divides the climates into five main groups and several types and subtypes. Each particular climate type is represented by a 2 to 4 letter symbol.

GROUP A: Tropical/megathermal climates

Tropical climates are characterized by constant high temperature — all twelve months of the year have average temperatures of 18 °C or higher. They are subdivided as follows:

- Tropical rainforest climate: All twelve months have average precipitation of at least 60 mm. These climates usually occur within 5–10° latitude of the equator. In some eastern-coast areas, they may extend to as much as 25° away from the equator. This climate is dominated by the Doldrums Low Pressure System all year round, and therefore has no natural seasons.
  - Examples:
    - Belém, Brazil
    - Hilo, Hawaii, United States
    - Singapore
  - Some of the places that have this climate are indeed uniformly and monotonously wet throughout the year, but in many cases the period of higher sun and longer days is distinctly wettest or the time of lower sun and shorter days may have more rain.
  - A few places with this climate are found at the outer edge of the tropics, almost exclusively in the Southern Hemisphere; one example is Santos, Brazil. Note. The term aseasonal refers to the lack in the tropical zone of large differences in day light hours and mean monthly temperature throughout the year. There are annual cyclic changes in the tropics, not as predictable as those in the temperate zone, albeit unrelated to temperature but to water availability whether as rain, mist, soil, or ground water. Plant response, animal, and human activities are tuned to this *seasonality*. Indeed, in tropical South America and Central America, the *rainy season* is called *Invierno* or *Inverno*, even though it could occur in the northern hemisphere summer; likewise, the *dry season* is called *Verano* or *Verão* and can occur in the northern hemisphere winter.
- Tropical monsoon climate: This type of climate, most common in southern Asia and West Africa, results from the monsoon winds which change direction according to the seasons. This climate has a driest month with rainfall less than 60 mm, but more than.
  - Examples:
    - Conakry, Guinea
    - Chittagong, Bangladesh
    - Miami, Florida, United States
    - Cairns, Australia
  - There is also another scenario under which some places fit into this category; this is referred to as the *trade-wind littoral* climate because easterly winds bring enough precipitation during the "winter" months to prevent the climate from becoming a tropical wet-and-dry climate. Nassau, Bahamas is included among these locations.

- Tropical wet and dry or savanna climate: These climates have a pronounced dry season, with the driest month having precipitation less than 60 mm and also less than.
  - Examples:
    - Jakarta, Indonesia
    - Mumbai, Maharashtra, India
    - Rio de Janeiro, Rio de Janeiro, Brazil
    - Veracruz, Veracruz, Mexico
    - Lagos, Lagos State, Nigeria
    - Darwin, Northern Territory, Australia
  - Most places that have this climate are found at the outer margins of the tropical zone, but occasionally an inner-tropical location also qualifies. Actually, the Caribbean coast, eastward from Urabá gulf on the Colombia–Panamá border to the Orinoco river delta, on the Atlantic ocean have long dry periods. This condition extends to the Lesser Antilles and Greater Antilles forming the Circumcaribbean dry belt. The length and severity of the dry season diminishes inland; at the latitude of the Amazon river — which flows eastward, just south of the equatorial line — the climate is *Af*. East from the Andes, between the dry, arid Caribbean and the ever-wet Amazon are the Orinoco river' Llanos or Savannas, from where this climate takes its name.
  - Sometimes *As* is used in place of *Aw* if the dry season occurs during the time of higher sun and longer days. This is the case in parts of Hawaii, East Africa and Sri Lanka, for instance. In most places that have tropical wet and dry climates, however, the dry season occurs during the time of lower sun and shorter days because of rainshadow effects during the 'high-sun' part of the year.
  - Examples:
    - Honolulu, Hawaii, United States

GROUP B: Dry climates

These climates are characterized by the fact that precipitation is less than potential evapotranspiration. The threshold is determined as follows:

- To find the precipitation threshold, multiply the average annual temperature in °C by 20, then add 280 if 70% or more of the total precipitation is in the high-sun half of the year or 140 if 30%–70% of the total precipitation is received during the applicable period, or 0 if less than 30% of the total precipitation is so received.
- If the annual precipitation is less than half the threshold for Group B, it is classified as *BW*; if it is less than the threshold but more than

half the threshold, it is classified as *BS*. If it's more than the threshold, the area does not have a Group B climate.

- A third letter can be included to indicate temperature. Originally, *h* signified low latitude climate while *k* signified middle latitude climate but the more common practice today is to use *h* to mean that the coldest month has an average temperature that is above 0 °C with *k* denoting that at least one month averages below 0 °C.
  - Examples:
    - Yuma, Arizona, United States (*BWh*)
    - Mexicali, Mexico (*BWh*)
    - Konya, Turkey (*BSk*)
    - Cobar, New South Wales, Australia (*BSh*)
    - Murcia, Spain (*BSh*)
    - Almeria, Spain (*BSh*)
    - Boise, Idaho, United States (*BSk*)
    - Denver, Colorado, United States (*BSk*)
    - Medicine Hat, Alberta, Canada (*BSk*)
    - Porto Santo, Portugal (*BSh*)
    - La Serena, Chile (*BSk*)
    - Mossoró, Rio Grande do Norte, Brazil (*BSh*)
    - Dubai, United Arab Emirates (*BWh*)
    - Jodhpur, Rajasthan, India (*BWh*)
    - Tijuana, Mexico (*BSh*)
  - Desert areas, situated along the west coasts of continents at tropical or near-tropical locations, are characterized by cooler temperatures than encountered elsewhere at comparable latitudes and frequent fog and low clouds, despite the fact that these places rank among the driest on earth in terms of actual precipitation received. This climate is sometimes labelled *BWn* and examples can be found at Lima, Peru and Walvis Bay, Namibia. The *BSn* category can be found in foggy coastal steppes.
- On occasion, a fourth letter is added to indicate if either the winter or summer is "wetter" than the other half of the year. To qualify, the wettest month must have at least 60 mm of average precipitation if all twelve months are above 18 °C, or 30 mm if not; plus at least 70% of the total precipitation must be in the same half of the year as the wettest month — but the letter used indicates when the *dry* season occurs, not the "wet" one. This would result in Khartoum, Sudan being reckoned as *BWhw*, Niamey, Niger as *BShw*, El Arish, Egypt as *BWhs*, Asbi'ah, Libya as *BShs*, Ömnögovi Province, Mongolia as *BWkw*, and Xining, China as *BSkw*. If the standards for neither *w* nor *s* are met, no fourth letter is added.

GROUP C: Temperate/mesothermal climates

These climates have an average temperature above 10 °C in their warmest months, and a coldest month average between –3 °C and 18 °C. Some climatologists, particularly in the United States, however, prefer to observe 0 °C rather than –3 °C in the coldest month as the boundary between this group and Group D; this is done to prevent certain headland locations in or near New England — principally Cape Cod — and such nearby islands as Nantucket and Martha's Vineyard, from fitting into the Maritime Temperate category noted below; this category is alternately known as the *Marine West Coast* climate, and eliminating the aforementioned locations indeed confines it exclusively to places found along the western margins of the continents, at least in the Northern Hemisphere. This also moves some mid-latitude areas – such as parts of the Ohio Valley and some areas in the Mid-Atlantic States, plus parts of east-central Asia – from humid subtropical to humid continental.

- The second letter indicates the precipitation pattern — *w* indicates dry winters, *s* indicates dry summers and *f* means significant precipitation in all seasons.
- The third letter indicates the degree of summer heat — *a* indicates warmest month average temperature above 22 °C with at least 4 months averaging above 10 °C, *b* indicates warmest month averaging below 22 °C, but with at least 4 months averaging above 10 °C, while *c* means 3 or fewer months with mean temperatures above 10 °C.
- The order of these two letters is sometimes reversed, especially by climatologists in the United States.
- Group C climates are subdivided as follows:
  - Dry-summer subtropical or Mediterranean climates: These climates usually occur on the western sides of continents between the latitudes of 30° and 45°. These climates are in the polar front region in winter, and thus have moderate temperatures and changeable, rainy weather. Summers are hot and dry, due to the domination of the subtropical high pressure systems, except in the immediate coastal areas, where summers are milder due to the nearby presence of cold ocean currents that may bring fog but prevent rain.
  - Examples:
  - Amman, Jordan (*Csa*)
  - Antalya, Turkey (*Csa*)
  - Athens, Greece (inland) (*Csa*)
  - Cape Town, South Africa (coastal) (*Csb*)
  - Cape Town, South Africa (inland) (*Csa*)
  - Coimbra, Portugal (*Csa*)

- Izmir, Turkey (*Csa*)
- Jerash, Jordan (*Csa*)
- Jerusalem, Israel (*Csa*)
- Lisbon, Portugal (*Csa*)
- Los Angeles, California, United States (inland) (*Csa*)
- Los Angeles, California, United States (coast) (*Csb*)
- Madrid, Spain (*Csa*)
- Malaga, Spain (*Csa*)
- Marseille, France (*Csa*)
- Medford, Oregon, United States (inland) (*Csa*)
- Palermo, Italy (*Csa*)
- Perth, Australia (*Csa*)
- Porto, Portugal (*Csb*)
- Risan, Montenegro (*Csb*)
- Sacramento, United States (*Csa*)
- San Francisco, California, United States (*Csb*)
- San Jose, California, United States (*Csb*)
- Sanremo, Italy (*Csa*)
- Santiago, Chile (*Csb*)
- Seville, Spain (*Csa*)
- Split, Croatia (*Csa*)
- Tel Aviv, Israel (*Csa*)

— Under the Köppen-Geiger classification, dry-summer subtropical extends to additional areas not typically associated with a typical Mediterranean climate, such as much of the Pacific Northwest, much of southern Chile, parts of west-central Argentina, and northern Spain and Portugal. Many of these areas would be Oceanic, except dry-summer patterns meet Köppen's *Cs* minimum thresholds. Additional highland areas in the subtropics also meet *Cs* requirements, although they too are not normally associated with Mediterranean climates.

- Examples:
- Portland, Oregon, United States
- Seattle, Washington, United States (*Csb* sometimes *Cfb*)
- Victoria, British Columbia, Canada (*Csb* sometimes *Cfb*)

— Humid subtropical climates: These climates usually occur in the interiors of continents, or on their east coasts, mainly in the high 20s and 30s latitude. Unlike the Mediterranean climates, the summers are humid due to unstable tropical air masses, or onshore Trade Winds. In eastern Asia, winters can be dry because of the Siberian high pressure system, and summers very wet due to the Southwest Asian monsoonal influence.

- Examples:

- Atlanta, United States (*Cfa — uniform precipitation distribution*)
- Belgrade, Serbia (*Cfa*)
- Varna, Bulgaria (*Cfa*)
- Odessa, Ukraine (*Cfa*)
- Tbilisi, Georgia (*Cfa*)
- Milan, Italy (*Cfa*)
- Sydney, Australia (*Cfa — uniform precipitation distribution*)
- Buenos Aires, Argentina (*Cfa — uniform precipitation distribution*)
- Porto Alegre, Brazil (*Cfa — uniform precipitation distribution*)
- Okinawa, Japan (*Cfa*)
- Brisbane, Queensland, Australia (*Cfa — uniform precipitation distribution*)
- Durban, KwaZulu-Natal, South Africa (*Cfa — uniform precipitation distribution*)
- Guadalajara, Jalisco, Mexico (*Cwa*)
- São Paulo, São Paulo, Brazil (*Cwa*)
- Lucknow, Uttar Pradesh, India (*Cwa*)
- Zhengzhou, China (*Cwa*)
- Hong Kong (*Cwa*)

— Maritime Temperate climates or Oceanic climates: *Cfb* climates usually occur on the western sides of continents between the latitudes of 45° and 55°; they are typically situated immediately poleward of the Mediterranean climates, although in Australia and extreme southern Africa this climate is found immediately poleward of the humid subtropical climate, and at a somewhat lower latitude. In western Europe, this climate occurs in coastal areas up to 63°N latitude. These climates are dominated all year round by the polar front, leading to changeable, often overcast weather. Summers are cool due to cloud cover, but winters are milder than other climates in similar latitudes.

- Examples:
- Limoges, France (*Cfb — uniform precipitation distribution*)
- Liverpool, United Kingdom (*Cfb — uniform precipitation distribution*)
- Hamburg, Germany (*Cfb — uniform precipitation distribution*)
- Bergen, Norway (*Cfb — uniform precipitation distribution*)
- George, South Africa (*Cfb — uniform precipitation distribution*)
- Melbourne, Victoria, Australia (*Cfb — uniform precipitation distribution*)
- Hobart, Tasmania, Australia (*Cfb — uniform precipitation distribution*)
- Wellington, New Zealand (*Cfb — uniform precipitation*

distribution)

- Vancouver, Canada (*Cfb — uniform precipitation distribution, but with drying trend in summer similar to Csb*)
- Prince Rupert, British Columbia, Canada (*Cfb — uniform precipitation distribution*)
- Castro, Chile (*Cfb — uniform precipitation distribution*)
- *Cfb* climates are also encountered at high elevations in certain subtropical and tropical areas, where the climate would be that of a subtropical/tropical rain forest if not for the altitude. These climates are called "Highlands". Examples include:
- Curitiba, Brazil (*Cfb — uniform precipitation distribution*)
- Boone, North Carolina, USA (*Cfb — uniform precipitation distribution*)
- Higher elevations of Tamil Nadu, India (*e.g.* Ooty)
- Crkvice, Montenegro (*Cfsb —perhumid Mediterranean mountain climate without summer dryness,* Crkvice on Orjen also holds Europe's precipitation record – averaging 4927 mm/m² 1931–1960).
- The temperate climate with dry winters (Cwb) is a type of climate characteristic of the highlands inside the tropics of Mexico, Peru, Bolivia, Madagascar, Zambia, Zimbabwe and South Africa but it is also found in central Argentina, outside the tropics. Winters are noticeable and dry and summers very rainy. In the tropics the rainy season is provoked by the tropical airmasses and the dry winters by subtropical high pressure. Temperate temperatures are the consequence of altitude which become cooler year-round.
- Examples:
- Cuzco, Peru
- Addis Ababa, Ethiopia
- Bogotá, Colombia
- Mexico City, Mexico
- Campos do Jordão, São Paulo, Brazil
- La Paz, Bolivia
- Johannesburg, Gauteng, South Africa
- Gangtok, Sikkim, India

— Maritime Subarctic climates or Subpolar Oceanic climates: These climates occur poleward of the Maritime Temperate climates, and are confined either to narrow coastal strips on the western poleward margins of the continents, or, especially in the Northern Hemisphere, to islands off such coasts.

- Examples:
- Monte Dinero, Argentina (*Cfc — uniform precipitation distribution*)

- Reykjavík, Iceland (*Cfc — uniform precipitation distribution*)
- Tórshavn, Faroe Islands (*Cfc — uniform precipitation distribution*)
- Bodø, Norway (*Cfc — uniform precipitation distribution*)

GROUP D: Continental/microthermal climate

These climates have an average temperature above 10 °C in their warmest months, and a coldest month average below "3 °C. These usually occur in the interiors of continents, or on their east coasts, north of 40° North latitude. In the Southern Hemisphere, Group D climates are extremely rare due to the smaller land masses in the middle latitudes and the almost complete absence of land between 40°-60° South latitude, existing only in some highland locations.

- The second and third letters are used as for Group C climates, while a third letter of *d* indicates 3 or fewer months with mean temperatures above 10 °C *and* a coldest month temperature below "38 °C.
- Group D climates are subdivided as follows:
  - Hot Summer Continental climates: *Dfa* climates usually occur in the high 30s and low 40s latitudes, with a qualifying average temperature in the warmest month of >22°C. In eastern Asia *Dwa* climates extend further south due to the influence of the Siberian high pressure system, which also causes winters here to be dry, and summers can be very wet because of monsoon circulation.
    - Examples:
    - Chicago, Illinois, United States (*Dfa — summer wetter than winter*)
    - Toronto, Ontario, Canada (*Dfa — summer somewhat wetter than winter*)
    - Santaquin, Utah, United States (*Dfa — summer drier than winter*)
    - Beijing, China (*Dwa*)
    - *Dsa* exists only at higher elevations adjacent to areas with hot summer Mediterranean (*Csa*) climates, such as Cambridge, Idaho and Saqqez in Iranian Kurdistan.
  - Warm Summer Continental or Hemiboreal climates (*Dfb, Dwb, Dsb*): *Dfb* and *Dwb* climates are immediately north of Hot Summer Continental climates, generally in the high 40s and low 50s in latitude in North America and Asia, and also in central and eastern Europe and Russia, between the Maritime Temperate and Continental Subarctic climates, where it extends up to high 50s and even lowest 60 degrees latitude.
    - Examples:

- Moncton, New Brunswick, Canada (*Dfb — uniform precipitation distribution*)
- Minsk, Belarus (*Dfb — summer wetter than winter*)
- Saskatoon, Saskatchewan, Canada (*Dfb; summer wetter than winter*)
- Revelstoke, British Columbia, Canada (*Dfb — summer drier than winter*)
- Fargo, North Dakota, United States (*Dfb — winter drier than summer*)
- Helsinki, Finland (*Dfb — winter somewhat drier than summer*)
- Växjö, Sweden (*Dfb — winter somewhat drier than summer*)
- Vladivostok, Russia (*Dwb*)
- *Dsb* arises from the same scenario as *Dsa,* but at even higher altitudes or higher latitudes, and chiefly in North America since here the Mediterranean climates extend further poleward than in Eurasia; Mazama, Washington is one such location.

— Continental Subarctic or Boreal (taiga) climates (*Dfc, Dwc, Dsc*): *Dfc* and *Dwc* climates occur poleward of the other Group D climates, mostly in the 50s and low 60s North latitude, although it might occur as far north as 70° latitude.

- Examples:
- Sept-Îles, Quebec, Canada (*Dfc — uniform precipitation distribution*)
- Anchorage, Alaska, United States (*Dfc — summer wetter than winter*)
- Mount Robson, British Columbia, Canada (*Dfc — summer drier than winter*)
- Murmansk, Russia (*Dfc – summer wetter than winter*)
- Irkutsk, Russia (*Dwc*)
- Kirkenes, Finnmark, Norway (*Dfc – summer wetter than winter*)
- Luleå, Sweden (*Dfc – uniform precipitation distribution*)
- *Dsc,* like *Dsa* and *Dsb,* is confined exclusively to highland locations near areas that have Mediterranean climates, and is the rarest of the three as a still higher altitude is needed to produce this climate. Two examples are Zubaèki kabao, Montenegro (*Dfsc* perhumid Mediterranean snow climate) and Galena Summit, Idaho.

— Continental Subarctic climates with extremely severe winters (*Dfd, Dwd*): These climates occur only in eastern Siberia. The names of some of the places that have this climate — most notably Verkhoyansk and Oymyakon — have become veritable synonyms for extreme, severe winter cold.

GROUP E: Polar climates

These climates are characterized by average temperatures below 10 °C in all twelve months of the year:

- Tundra climate: Warmest month has an average temperature between 0 °C and 10 °C. These climates occur on the northern edges of the North American and Eurasian landmasses, and on nearby islands.
  - Examples:
    - Iqaluit, Nunavut, Canada
    - Provideniya, Russia
    - Nuuk, Greenland
    - Vardø, Norway
    - Longyearbyen, Svalbard
    - Barrow, Alaska
    - *ET* is also found at high elevations outside the polar regions, above the tree line — as at Mount Washington, New Hampshire and Jotunheimen, Norway.
- Ice Cap climate: All twelve months have average temperatures below 0 °C. This climate is dominant in Antarctica and in inner Greenland.
- Occasionally, a third, lower-case letter is added to *ET* climates if either the summer or winter is clearly drier than the other half of the year; thus Herschel Island off the coast of Canada's Yukon Territory, becomes *ETw,* with Pic du Midi de Bigorre in the French Pyrenees acquiring an *ETs* designation. If the precipitation is more or less evenly spread throughout the year, *ETf* may be used, such as for Hebron, Labrador. When the option to include this letter is exercised, the same standards that are used for Groups C and D apply, with the additional requirement that the wettest month must have an average of at least 30 mm precipitation. Seasonal precipitation letters are almost never attached to *EF* climates, mainly due to the difficulty in distinguishing between falling and blowing snow, as snow is the sole source of moisture in these climates.

GROUP H: Alpine climates

The Alpine climates are considered to be part of group E.

- Alpine Climate
  - Examples:
    - The Cascade Mountains
    - The Rocky Mountains
    - The Andes
    - The Himalayas
    - The Tibetan Plateau

- The Eastern Highlands of Africa
- The Snowy Mountains of Australia
- The West Coast Range of Tasmania
- The Alps

Trewartha climate classification scheme

The Trewartha climate classification scheme is a modified version of the Köppen system. It attempts to redefine the broad climatic groups in such a way as to be closer to vegetational zoning, especially in the United States. Under the standard Köppen system western Washington and Oregon are classed into the same climate as southern California, even though the two regions have strikingly different weather and vegetation. It also classes southern New England into the same climate as the Gulf Coast. Trewartha's modifications sought to reclass the Pacific Northwest seaboard as a different climate from California, and New England from the Gulf Coast.

- Group A: This the tropical climate realm, defined the same as in Köppen's scheme. Climates with no more than two dry months are classified *Ar*, while others are classified *Aw* if the dry season is at the time of low sun/short days or *As* if the dry season is at the time of high sun/long days. There was no specific monsoon climate identifier in the original scheme, but *Am* was added later, with the same parameters as Köppen's.
- Group B: *BW* and *BS* mean the same as in the Köppen scheme, with the Köppen *BWn* climate sometimes being designated *BM*. However, a different formula is used to quantify the aridity threshold: 10 X (*T* " 10) + 3*P*, with *T* equalling the mean annual temperature in degrees Celsius and *P* denoting the percentage of total precipitation received in the six high-sun months. If the precipition for a given location is less than the above formula, its climate is said to be that of a desert; if it is equal to or greater than the above formula but less than twice that amount, the climate is classified as steppe; and if the precipitation is more than double the value of the formula the climate is not in Group B. Unlike in Köppen's scheme, no thermal subsets exist within this group in Trewartha's, unless the Universal Thermal Scale is used.
- Group C: In the Trewartha scheme this category encompasses subtropical realm climates only. *Cs* and *Cw* have the same meanings as they do in Köppen's scheme, but the subtropical climate with no distinct dry season is designated *Cr* instead of Köppen's *Cf*.
- Group D: This group represents cold continental realms. Maritime temperate climates are denoted *DO* in the Trewartha classification, while continental climates are represented as *DCa*. For the continental climates, sometimes the third letter is omitted and *DC*

is simply used instead, and occasionally a precipitational seasonality letter is added to both the maritime and continental climates. The dividing point between the maritime and continental climates is 0 °C in the coldest month, rather than the Köppen value of "3 °C.

- Group E: This represents Ice realms, defined the same as in Köppen's scheme. In the original scheme, this group was not further divided; later, the designations *EO* and *EC* were created, with *EO* signifying that the coldest month averages above "10 °C, while *EC* means that at least one month has an average temperature of "10 °C or below. As in Group D, a third letter can be added to indicate seasonality of precipitation. There is no separate counterpart to the Köppen *Dfd/Dwd* climate in Trewartha's scheme.
- Group F: This is the polar climate group, split into *FT* and *FI*.
- Group H: Highland climates, in which altitude plays a role in determining climate classification. Specifically, this would apply if correcting the average temperature of each month to a sea-level value using the formula of adding 5.6 °C for each 1,000 meters of elevation would result in the climate fitting into a different thermal group than that into which the actual monthly temperatures place it. Sometimes *G* is used instead of *H* if the above is true and the altitude is 500 meters or higher but lower than 2,500 meters; but the *G* or *H* is placed in front of the applicable thermal letter rather than replacing it — and the second letter used reflects the *corrected* monthly temperatures, not the actual monthly temperatures.
- Universal Thermal Scale: An option exists to include information on both the warmest and coldest months for every climate by adding a third and fourth letter, respectively. The letters used conform to the following scale:
  - *i* — severely hot: Mean monthly temperature 35 °C (95 °F) or higher
  - *h* — very hot: 28 to 34.9 °C (82 to 95 °F)
  - *a* — hot: 23 to 27.9 °C (73 to 82 °F)
  - *b* — warm: 18 to 22.9 °C (64 to 73 °F)
  - *l* — mild: 10 to 17.9 °C (50 to 64 °F)
  - *k* — cool: 0.1 to 9.9 °C (32 to 50 °F)
  - *o* — cold: -9.9 to 0 °C (14 to 32 °F)
  - *c* — very cold: –24.9 to -10 °C (-13 to 14 °F)
  - *d* — severely cold: -39.9 to -25 °C (-40 to -13 °F)
  - *e* — excessively cold: –40 °C (–40 °F) or below.

Examples of the resulting designations include *Afaa* for Surabaya, Indonesia, *BWhl* for Aswan, Egypt, *Crhk* for Dallas, Texas, *DOlk* for London, *EClc* for Arkhangelsk, Russia, and *FTkd* for Barrow, Alaska.

Criticisms of the Köppen scheme

Some climatologists have argued that Köppen's system could be improved upon. One of the most frequently-raised objections concerns the temperate Group C category, regarded by many as overbroad. In *Applied Climatology*, John F. Griffiths proposed a new subtropical zone, encompassing those areas with a coldest month of between 6 °C and 18 °C, effectively subdividing Group C into two nearly equal parts.

Another point of contention involves the dry *B* climates; the argument here is that their separation by Köppen into only two thermal subsets is inadequate. Those who hold this view have suggested that the dry climates be placed on the same temperature continuum as other climates, with the thermal letter being followed by an additional capital letter — *S* for steppe or *W* for desert — as applicable.

A third idea is to create a *maritime polar* or *EM* zone within Group E to separate relatively mild marine locations from the colder, continental tundra climates. Specific proposals vary; some advocate setting a coldest-month parameter, such as "7 °C, while others support assigning the new designation to areas with an average annual temperature of above 0 °C.

The accuracy of the 10 °C warmest-month line as the start of the polar climates has also been questioned; Otto Nordenskiöld, for example, devised an alternate formula: $W = 9$ " $0.1\ C$, with W representing the average temperature of the warmest month and C that of the coldest month, both in degrees Celsius. This boundary does appear to more closely follow the tree line, or the latitude poleward of which trees cannot grow, than the 10 °C warmest-month isotherm; the former tends to run poleward of the latter near the western margins of the continents, but at a lower latitude in the landmass interiors, the two lines crossing at or near the east coasts of both Asia and North America.

World Map of the Köppen–Geiger climate classification for the period 1951–2000

Based on recent data sets from the Climatic Research Unit of the University of East Anglia and the Global Precipitation Climatology Centre at the German Weather Service, a new digital Köppen–Geiger world map on climate classification for the second half of the 20th century has been compiled.

THORNTHWAITE'S CLIMATES CLASSIFICATION

Salient features of 1931 classification. - This classification is almost similar to Koppen's classification in so far as it defines climatic boundaries quantitatively, and it is also based on vegetation. Like Koppen's scheme it also makes. use of letter combinations to designate climatic types. However, it differs from Koppen's classification on two scores: first, he introduced an expression for precipitation efficiency, and second, he made use of an index of thermal efficiency. Thornthwaite makes moisture the primary classificatory factor for a T/E index of over 31.

Since Thornthwaite adopted the precipitation effectiveness and temperature efficiency indices for his climate' classification, the delimitation of the climatic boundaries becomes difficult and vague. Besides, under this classification there are almost 32 major climatic types shown on the world map-this is almost three times greater than Koppen's climatic types. But the point to be remembered is that in Thornthwaite's classification the letter symbols used are relatively less in number than those in Koppen's classification.

Lastly, it may be pointed out that the climatic classifications as devised by Koppen and Thornthwaite are more useful and appealing to zoologists, botanists, and geographers. But these schemes of classification are not so useful for meteorologists and climatologists because the interplay between the weather elements and other climatic factors is not clearly shown.

THORNTHWAITE'S 1948 CLASSIFICATION

In 1948 Thornthwaite proposed a new classification of climate which is his most important contribution. His second classification is based. on the concept of potential evapotranspiration, which represents the amount of moisture that would be transferred to the atmosphere by evaporation of liquid or solid water plus transpiration from living tissues, principally plants if it were available. The potential evapotranspiration is calculated from the mean monthly temperature, with corrections for day length. For a 30-day month:

- PE (in cm) = 1.6 (10t/1)a
- Where I = the sum of 12 months of (t/5)1.514
- a = a further complex function of I.

The monthly water surplus (S) or deficit (D) is calculated from a moisture budget assessment including stored soil moisture. A moisture index is given by the following formula:

- IM = (100S-60D)PE

The most characteristic feature of this classification scheme is that the temperature efficiency is calculated from the PE value, this being a function of temperature. Using computed indices of moisture and heat, Thornthwaite defined the moisture and thermal provinces.

CRITICISM OF THORNTHWAITE'S 1948 CLASSIFICATION

This scheme of climatic classification has been successfully applied to many regions, but because of certain limitations inherent in it no world map has yet been prepared. This classification system has proved most satisfactory in case of North America where vegetation boundaries nearly coincide with particular PE values. But it is not satisfactory for the tropical and semi-arid areas. However, Thornthwaite does not determine his climatic boundaries on the basis of vegetation boundaries. In this respect, he differs from Koppen in case of his second classification. This classification is quantitative as well as empirical, but it does not take into account various causative factors of climate.

Lastly, Thornthwaite as a climatologist developed the concept of the soil-moisture balance, and he preferred to use it as the foundation of his climate system. He was convinced that the soil-moisture balance represents availability of moisture for plants, and an assessment of the availability of surplus moisture to supply stream flow and ground water. This concept implies that precipitation alone does not indicate the amount of water actually available to plants. The amount and availability of soil moisture is also affected, besides other factors, by the losses due to evapotranspiration.

The most significant contribution of Thornthwaite has been that the concept of evapotranspiration has been applied in practical studies of the water balance as regards the problems of water use. However, in spite of numerous plus points of Thornthwaite's classification of the climate, his system is seldom used in an introductory course because it involves a lot of very complicated calculations.

HYDROLOGICAL CYCLE

Most of the earth's water sources get their water supplies from precipitation, which may fall in various forms, such as, rain, snow, hail, dew etc. Rains no doubt, form the principal and the major part of the resultant supplies. When rain starts falling, it is first of all intercepted by buildings and other objects. When the rainfall rate exceeds the interception rate, water starts reaching the ground and infiltration starts. This is the source of groundwater storage.

SOURCES OF WATER

The primary sources of water include: rainwater, surface water and groundwater. The distribution of water, however, is quite varied; many locations have plenty of it while others have very little. Water exists on earth in three forms solid, liquid or gas. Oceans, rivers, clouds, and rain, all of which contain water, are in a frequent state of change. However, the total amount of the earth's water does not change. Owing to glaciers, rivers and groundwater flow.

Water is essential to life. Without it, the biosphere that exists on the surface of the earth would not be possible. The earth is called as the 'water' planet, water's molecular arrangement of water is very simple, two hydrogen atoms to each oxygen atom. One special characteristic of water is its ability to change state very easily under earth conditions. It can be found readily on the planet in all of its three forms, solid, liquid, and gas.

The average annual rainfall in the country is 1170 mm, which corresponds to annual precipitation, including snowfall of 4000 Billion Cubic Metres. Out of this volume of precipitation, only 1869 BCM appears as average annual potential flow in rivers. Due to various constraints, only 1123 BCM is assessed as the average annual utilisable water – 690 BCM from surface water and 433 BCM from groundwater.

The present total water use is 634 BCM of which 83% is for irrigation. This is projected to grow to 813 BCM by 2010, 1093 BCM by 2025 and 1447 BCM by 2050, against utilizable quantum of 1123 BCM. Thus the demand will outstrip availability in another 35 to 40 years. The Central Ground Water Board has estimated the present annual groundwater draft as 231 BCM.

HYDROLOGIC CYCLE

The movement of water on the earth's surface and through the atmosphere is known as the hydrologic cycle. Water is taken up by the atmosphere from the earth's surface in vapour form through evaporation. It may then be moved from place to place by the wind until it is condensed back to its liquid phase to form clouds. Water then returns to the surface of the earth in the form of either liquid or solid precipitation. Water transport can also take place on or below the earth's surface by flow.

The hydrologic cycle is used to model the storage and movement of water between the biosphere, atmosphere, lithosphere and hydrosphere. Water is stored in the following reservoirs: atmosphere, oceans, lakes, rivers, glaciers, soils, snowfields, and groundwater. It moves from one reservoir to another by processes like: evaporation, condensation, precipitation, deposition, runoff, infiltration, sublimation, transpiration, and groundwater flow.

Water is stored in the atmosphere in all three states of matter. Water vapour in the atmosphere is commonly referred to as humidity. If liquid and solid forms of water can overcome atmospheric updrafts they can fall to the Earth's surface as precipitation. The formation of ice crystals and water droplets occurs when the atmosphere is cooled to a temperature that causes condensation or deposition. Four processes that can cause atmospheric cooling are: orographic uplift; convectional uplift; air mass convergence; and radiative energy loss.

Precipitation can be defined as any aqueous deposit, in liquid or solid form, that develops in a saturated atmospheric environment and generally falls from clouds. A number of different precipitation types have been classified by meteorologists including rain, freezing rain, snow, ice pellets, snow pellets, and hail. Fog represents the saturation of air near the ground surface. Classification of fog types is accomplished by the identification of the mechanism that caused the air to become saturated. The distribution of precipitation on the Earth's surface is generally controlled by the absence or presence of mechanisms that lift air masses to cause saturation. It is also controlled by the amount of water vapour held in the air, which is a function of air temperature. In certain locations on the Earth, acid pollutants from the atmosphere are being deposited in dry and wet forms to the Earth's surface. Scientists generally call this process acid deposition. If the deposit is wet it can also be called acid precipitation. Normally, rain is slightly acidic. Acid precipitation, however, can have a pH as low as 2.3. Evaporation and transpiration are the two processes that move water from the Earth's surface to its atmosphere. Evaporation is movement of free water to the atmosphere as a gas. It requires large amounts of energy. Transpiration is the movement of water through a plant to the atmosphere. Scientists use the term evapotranspiration to describe both processes.

In general, the following four factors control the amount of water entering the atmosphere via these two processes: energy availability; the humidity gradient away from the evaporating surface; the wind speed immediately above the surface; and water availability. Agricultural scientists sometimes refer to two types of evapotranspiration: Actual Evapotranspiration and Potential Evapotranspiration. The growth of crops is a function of water supply. If crops experience drought, yields are reduced. Irrigation can supply crops with supplemental water. By determining both actual evapotranspiration and potential evapotranspiration a farmer can calculate the irrigation water needs of their crops.

The distribution of precipitation falling on the ground surface can be modified by the presence of vegetation. Vegetation in general, changes this distribution because of the fact that it intercepts some the falling rain. How much is intercepted is a function of the branching structure and leaf density of the vegetation. Some of the water that is intercepted never makes it to the ground surface. Instead, it evaporates from the vegetation surface directly back to the atmosphere. A portion of the intercepted water can travel from the leaves to the branches and then flow down to the ground via the plant's stem. This phenomenon is called stem flow. Another portion of the precipitation may flow along the edge of the plant canopy to cause canopy drip. Both of the processes described above can increase the concentration of the water added to the soil at the base of the stem and around the edge of the plant's canopy.

Rain that falls through the vegetation, without being intercepted, is called through fall. Infiltration is the movement of water from precipitation into the soil layer. Infiltration varies both spatially and temporally due to a number of environmental factors. After a rain, infiltration can create a condition where the soil is completely full of water. This condition is, however, only short-lived as a portion of this water quickly drains via the force exerted on the water by gravity. The portion that remains is called the field capacity. In the soil, field capacity represents a film of water coating all individual soil particles to a thickness of 0.06 mm. The soil water from 0.0002 to 0.06 mm can be removed from the soil through the processes of evaporation and transpiration. Both of these processes operate at the surface. Capillary action moves water from one area in the soil to replace losses in another area. This movement of water by capillary action generally creates a homogeneous concentration of water throughout the soil profile. Losses of water stop when the film of water around soil particles reaches 0.0002 mm. Water held from the surface of the soil particles to 0.0002 mm is essentially immobile and can only be completely removed with high temperatures. Within the soil system, several different forces influence the storage of water.

Runoff is the surface flow of water to areas of lower elevation. On the microscale, runoff can be seen as a series of related events. At the global scale runoff flows from the landmasses to the oceans. The Earth's continents experience runoff because of the imbalance between precipitation and evaporation.

Through flow is the horizontal subsurface movement of water on continents. Rates of through flow vary with soil type, slope gradient, and the concentration of water in the soil. Groundwater is the zone in the ground that is permanently saturated with water. The top of groundwater is known as the water table. Groundwater also flows because of gravity to surface basins of water located at lower elevations. The flow of water through a stream channel is commonly called stream flow or stream discharge. On many streams humans gauge stream flow because of the hazards that can result from too little or too much flow. Mechanical gauging devices record this information on a graph known as a hydrograph. In the online notes there is a representation of a hydrograph showing some of its typical features.

Oceans cover most of the Earth's surface. On average, the depth of the world's oceans is about 3.9 kilometers. However, maximum depths can be greater than 11 kilometers. The distribution of land and ocean surfaces on the Earth is not homogeneous. In the Southern Hemisphere there is 4 times more ocean than land. Ratio between land and ocean is almost equal in the Northern Hemisphere.

The water found in the ocean is primarily a by product of the lithospheric solidification of rock that occurred early in the Earth's history. A second source of water is volcanic eruptions. The dissolved constituents found in the ocean come from the transport of terrestrial salts in weathered sediments by leaching and stream runoff. Seawater is a mixture of water and various salts. Chlorine, sodium, magnesium, calcium, potassium, and sulfur account for 99 % of the salts in seawater. The presence of salt in seawater allows ice to float on top of it. Seawater also contains small quantities of dissolved gases including: carbon dioxide, oxygen, and nitrogen. These gases enter the ocean from the atmosphere and from a variety of organic processes. Seawater changes its density with variations in temperature, salinity, and ocean depth. Seawater is least dense when it is frozen at the ocean surface and contains no salts.

Highest seawater densities occur at the ocean floor. Atmospheric circulation drives the movement of ocean currents. Within each of the ocean, the patterns of these currents are very similar. In each basin, the ocean currents form several closed circulation patterns known as gyres. A large gyre develops at the subtropics centered at about 30 degrees of latitude in the Southern and Northern Hemisphere. In the Northern Hemisphere, several smaller gyres develop with a center of rotation at 50 degrees. Similar patterns do not develop in the middle latitudes of the Southern Hemisphere. In this area, ocean currents are not bound by continental masses. Ocean currents differ from each other by direction of flow, by speed of flow, and by relative temperature.

Approximately 97 % of all the water on the Earth is in the oceans. The other 3 % is held as freshwater in glaciers and icecaps, groundwater, lakes, soil, the atmosphere, and within life. Water is continually cycled between its various reservoirs. This cycling occurs through the processes of evaporation, condensation, precipitation, deposition, runoff, infiltration, sublimation, transpiration, melting, and groundwater flow.

Hydrology, as discussed earlier, is the study of the movement and distribution of water throughout the Earth, and thus addresses both the hydrologic cycle and water resources. Many processes work together to keep Earth's water moving in a cycle. There are five processes at work in the hydrologic cycle: condensation, precipitation, infiltration, runoff, and evapotranspiration. These occur simultaneously and, except for precipitation, continuously. Water vapour condenses to form clouds, which result in precipitation when the conditions are suitable. Precipitation falls to the surface and infiltrates the soil or flows to the ocean as runoff. Surface water, evaporates, returning moisture to the atmosphere, while plants return water to the atmosphere by transpiration. The water cycle - technically known as the hydrologic cycle - is the circulation of water within the earth's hydrosphere, involving changes in the physical state of water between liquid, solid, and gas phases. The hydrologic cycle refers to the continuous exchange of water between atmosphere, land, surface and subsurface waters, and organisms. In addition to storage in various compartments; the multiple cycles that make up the earth's water cycle involve five main physical actions: *evaporation, precipitation, infiltration, runoff, and subsurface flow*

Evaporation

It occurs when radiant energy from the sun heats water, causing the water molecules to become so active that some of them rise into the atmosphere as vapour. It is the transfer of water from bodies of surface water into the atmosphere. This transfer entails a change in the physical nature of water from liquid to gaseous phases. Along with evaporation can be counted transpiration from plants. Thus, this transfer is sometimes referred to as evapotranspiration. About 90% of atmospheric water comes from evaporation, while the remaining

10% is from transpiration. Transpiration occurs when plants take in water through the roots and release it through the leaves, a process that can clean water by removing contaminants and pollution. Evapotranspiration is water *evaporating* from the ground and *transpiration* by plants. Evapotranspiration is also the way water vapour re-enters the atmosphere.

Precipitation

In cold air way up in the sky, rain clouds will often form. Rising warm air carries water vapor high into the sky where it cools, forming water droplets around tiny bits of dust in the air. Some vapor freezes into tiny ice crystals which attract cooled water drops. The drops freeze to the ice crystals, forming larger crystals we call snowflakes. When the snowflakes become heavy, they fall. When the snowflakes meet warmer air on the way down, they melt into raindrops. In tropical climates, cloud droplets combine together around dust or sea salt particles. They bang together and grow in size until they're heavy enough to fall.

Sometimes there is a layer of air in the clouds that is above freezing, or 32° F. Then closer to the ground the air temperature is once again below freezing. Snowflakes partially melt in the layer of warmer air, but then freeze again in the cold air near the ground. This kind of precipitation is called sleet. It bounces when it hits the ground. If snowflakes completely melt in the warmer air, but temperatures are below freezing near the ground, rain may freeze on contact with the ground or the streets. This is called freezing rain, and a significant freezing rain is called an ice storm. Ice storms are extremely dangerous because the layer of ice on the streets can cause traffic accidents. Ice can also build up on tree branches and power lines, causing them to break and our lights to go out. There is another kind of precipitation that comes from thunderstorms called hail.

Infiltration

Under some circumstances precipitation actually evaporates before it reaches the surface. More often, though, precipitation reaches the Earth's surface, adding to the surface water in streams and lakes, or infiltrating the A portion of the precipitation that reaches the Earth's surface seeps into the ground through the process called infiltration. Infiltration into the ground is the transition from surface water to groundwater. The infiltration rate will depend upon soil or rock permeability as well as other factors. Infiltrated water may reach another compartment known as groundwater. Groundwater tend to move slowly, so the water may return as surface water after storage within an aquifer for a period of time that can amount to thousands of years in some cases. Water returns to the land surface at lower elevation than where it infiltrated, under the force of gravity or gravity induced pressures.

Runoff

The amount of water that infiltrates the soil varies with the degree of land slope, the amount and type of vegetation, soil type and rock type, and whether the soil is already saturated by water. The more openings in the surface, the more infiltration occurs. Water that doesn't infiltrate the soil flows on the surface as runoff.

Precipitation that reaches the surface of the Earth but does not infiltrate the soil is called runoff. Runoff can also come from melted snow and ice. Also it includes the variety of ways by which land surface water moves down slope to the oceans. Water flowing in streams and rivers may be delayed for a time in lakes. Not all precipitated water returns to the sea as runoff; much of it evaporates before reaching the ocean or reaching an aquifer.

Subsurface Flow

Surface flow incorporates movement of water within the earth, either within the recharge zone or aquifers. After infiltrating, subsurface water may return to the surface or eventually seep into the ocean.

LET US SUM UP

The main sources of water include rainwater, surface water and groundwater. The distribution of water varies as per its locations. Water also exists on earth as a solid, liquid or gas, oceans, rivers, clouds, and rain, all of which contain water, are in a frequent state of change. Hydrologic Cycle recycles the earth's valuable water supply. The sun is the energy that powers this remarkable process. Its energy in the form of light, and heat causes water to evaporate from oceans, rivers, lakes and even puddles. Warm air currents rising from the earth's surface lift this water vapour up into the atmosphere. When the air currents reach the cooler layers of the atmosphere, the water vapour condenses around and clings on to fine particles in the air. This step is called condensation. When enough vapours attach itself to tiny pieces of dust, pollen or pollutants, it forms a cloud. As the air gets more and more moist, the droplets that form the clouds grow larger and larger. Eventually they will get so big that the swirling atmospheric winds can no longer hold them up. The droplets then fall from the sky as precipitation. Precipitation can be in the form of rain, snow, sleet or hail depending on other atmospheric conditions such as temperature. Some of the precipitation will be absorbed into the ground. This is called infiltration. Once in the ground, the water can join the earth's groundwater supply. This is one of the world's largest storehouses of water. The water could also be absorbed from the ground by the roots of plants.

CLIMATE CHANGE

Climate change is a long-term change in the statistical distribution of weather patterns over periods of time that range from decades to millions of years. It may be a change in the average weather conditions or a change in the distribution of weather events with respect to an average, for example, greater or fewer extreme weather events. Climate change may be limited to a specific region, or may occur across the whole Earth.

In recent usage, especially in the context of environmental policy, climate change usually refers to changes in modern climate. It may be qualified as anthropogenic climate change, more generally known as global warming or anthropogenic global warming.

TERMINOLOGY

The most general definition of *climate change* is a change in the statistical properties of the climate system when considered over periods of decades or longer, regardless of cause. Accordingly, fluctuations on periods shorter than a few decades, such as El Niño, do not represent climate change.

The term sometimes is used to refer specifically to climate change caused by human activity; for example, the United Nations Framework Convention on Climate Change defines climate change as "a change of climate which is attributed directly or indirectly to human activity that alters the composition of the global atmosphere and which is in addition to natural climate variability observed over comparable time periods." In the latter sense climate change is synonymous with global warming.

CAUSES

Factors that can shape climate are climate forcings. These include such processes as variations in solar radiation, deviations in the Earth's orbit, mountain-building and continental drift, and changes in greenhouse gas concentrations. There are a variety of climate change feedbacks that can either amplify or diminish the initial forcing. Some parts of the climate system, such as the oceans and ice caps, respond slowly in reaction to climate forcing because of their large mass. Therefore, the climate system can take centuries or longer to fully respond to new external forcings.

Human influences

In the context of climate variation, anthropogenic factors are human activities which affect the climate. The scientific consensus on climate change is, "that climate is changing and that these changes are in large part caused by human activities."

"Science has made enormous inroads in understanding climate change and its causes, and is beginning to help develop a strong understanding of current and potential impacts that will affect people today and in coming decades. This understanding is crucial because it allows decision makers to place climate change in the context of other large challenges facing the nation and the world. There are still some uncertainties, and there always will be in understanding a complex system like Earth's climate. Nevertheless, there is a strong, credible body of evidence, based on multiple lines of research, documenting that climate is changing and that these changes are in large part caused by human activities. While much remains to be learned, the core phenomenon, scientific questions, and hypotheses have been examined thoroughly and have stood firm in the face of serious scientific debate and careful evaluation of alternative explanations." — *United States National Research Council, Advancing the Science of Climate Change*

Consequently, the debate is shifting onto ways to reduce further human impact and to find ways to adapt to change that has already occurred and is anticipated to occur in the future.

Of most concern in these anthropogenic factors is the increase in CO_2 levels due to emissions from fossil fuel combustion, followed by aerosols and cement manufacture. Other factors, including land use, ozone depletion, animal agriculture and deforestation, are also of concern in the roles they play - both separately and in conjunction with other factors - in affecting climate, microclimate, and measures of climate variables.

Plate tectonics

Over the course of millions of years, the motion of tectonic plates reconfigures global land and ocean areas and generates topography. This can affect both global and local patterns of climate and atmosphere-ocean circulation.

The position of the continents determines the geometry of the oceans and therefore influences patterns of ocean circulation. The locations of the seas are important in controlling the transfer of heat and moisture across the globe, and therefore, in determining global climate. A recent example of tectonic control on ocean circulation is the formation of the Isthmus of Panama about 5 million years ago, which shut off direct mixing between the Atlantic and Pacific Oceans. This strongly affected the ocean dynamics of what is now the Gulf Stream and may have led to Northern Hemisphere ice cover. During the Carboniferous period, about 300 to 360 million years ago, plate tectonics may have triggered large-scale storage of carbon and increased glaciation. Geologic evidence points to a "megamonsoonal" circulation pattern during the time of the supercontinent Pangaea, and climate modeling suggests that the existence of the supercontinent was conducive to the establishment of monsoons.

The size of continents is also important. Because of the stabilizing effect of the oceans on temperature, yearly temperature variations are generally lower in coastal areas than they are inland. A larger supercontinent will therefore have more area in which climate is strongly seasonal than will several smaller continents or islands.

Solar output

The sun is the predominant source for energy input to the Earth. Both long- and short-term variations in solar intensity are known to affect global climate. Three to four billion years ago the sun emitted only 70% as much power as it does today. If the atmospheric composition had been the same as today, liquid water should not have existed on Earth.

However, there is evidence for the presence of water on the early Earth, in the Hadean and Archean eons, leading to what is known as the faint young sun paradox. Hypothesized solutions to this paradox include a vastly different atmosphere, with much higher concentrations of greenhouse gases than currently exist Over the following approximately 4 billion years, the energy output of the sun increased and atmospheric composition changed, with the oxygenation of the atmosphere around 2.4 billion years ago being the most notable alteration.

These changes in luminosity, and the sun's ultimate death as it becomes a red giant and then a white dwarf, will have large effects on climate, with the red giant phase possibly ending life on Earth. Solar output also varies on shorter time scales, including the 11-year solar cycle and longer-term modulations. Solar intensity variations are considered to have been influential in triggering the Little Ice Age, and some of the warming observed from 1900 to 1950.

The cyclical nature of the sun's energy output is not yet fully understood; it differs from the very slow change that is happening within the sun as it ages and evolves. While most research indicates solar variability has induced a small cooling effect from 1750 to the present, a few studies point toward solar radiation increases from cyclical sunspot activity affecting global warming. Interestingly, a 2010 study *suggests*, "that the effects of solar variability on temperature throughout the atmosphere may be contrary to current expectations."

Orbital variations

Slight variations in Earth's orbit lead to changes in the seasonal distribution of sunlight reaching the Earth's surface and how it is distributed across the globe. There is very little change to the area-averaged annually averaged sunshine; but there can be strong changes in the geographical and seasonal distribution.

The three types of orbital variations are variations in Earth's eccentricity, changes in the tilt angle of Earth's axis of rotation, and precession of Earth's axis. Combined together, these produce Milankovitch cycles which have a large impact on climate and are notable for their correlation to glacial and interglacial periods, their correlation with the advance and retreat of the Sahara, and for their appearance in the stratigraphic record.

Volcanism

Volcanism is a process of conveying material from the crust and mantle of the Earth to its surface. Volcanic eruptions, geysers, and hot springs, are examples of volcanic processes which release gases and/or particulates into the atmosphere.

Eruptions large enough to affect climate occur on average several times per century, and cause cooling for a period of a few years. The eruption of Mount Pinatubo in 1991, the second largest terrestrial eruption of the 20th century affected the climate substantially. Global temperatures decreased by about 0.5 °C. The eruption of Mount Tambora in 1815 caused the Year Without a Summer. Much larger eruptions, known as large igneous provinces, occur only a few times every hundred million years, but may cause global warming and mass extinctions.

Volcanoes are also part of the extended carbon cycle. Over very long time periods, they release carbon dioxide from the Earth's crust and mantle, counteracting the uptake by sedimentary rocks and other geological carbon dioxide sinks. According to the US Geological Survey, however, estimates are that human activities generate 100-300 times the amount of carbon dioxide emitted by volcanoes.

Ocean variability

The ocean is a fundamental part of the climate system. Short-term fluctuations such as the El Niño-Southern Oscillation, the Pacific decadal oscillation, the North Atlantic oscillation, and the Arctic oscillation, represent climate variability rather than climate change. On longer time scales, alterations to ocean processes such as thermohaline circulation play a key role in redistributing heat by carrying out a very slow and extremely deep movement of water, and the long-term redistribution of heat in the world's oceans.

PHYSICAL EVIDENCE FOR CLIMATIC CHANGE

Evidence for climatic change is taken from a variety of sources that can be used to reconstruct past climates. Reasonably complete global records of surface temperature are available beginning from the mid-late 19th century. For earlier periods, most of the evidence is indirect—climatic changes are inferred from changes in proxies, indicators that reflect climate, such as vegetation, ice cores, dendrochronology, sea level change, and glacial geology.

Historical and archaeological evidence

Climate change in the recent past may be detected by corresponding changes in settlement and agricultural patterns. Archaeological evidence, oral history and historical documents can offer insights into past changes in the climate. Climate change effects have been linked to the collapse of various civilisations.

Glaciers

Glaciers are considered among the most sensitive indicators of climate change. Their size is determined by a mass balance between snow input and melt output. As temperatures warm, glaciers retreat unless snow precipitation increases to make up for the additional melt; the converse is also true. Glaciers grow and shrink due both to natural variability and external forcings. Variability in temperature, precipitation, and englacial and subglacial hydrology can strongly determine the evolution of a glacier in a particular season. Therefore, one must average over a decadal or longer time-scale and/ or over a many individual glaciers to smooth out the local short-term variability and obtain a glacier history that is related to climate.

A world glacier inventory has been compiled since the 1970s, initially based mainly on aerial photographs and maps but now relying more on satellites. This compilation tracks more than 100,000 glaciers covering a total area of approximately 240,000 km^2, and preliminary estimates indicate that the remaining ice cover is around 445,000 km^2. The World Glacier Monitoring Service collects data annually on glacier retreat and glacier mass balance From this data, glaciers worldwide have been found to be shrinking significantly, with strong glacier retreats in the 1940s, stable or growing conditions during the 1920s and 1970s, and again retreating from the mid 1980s to present.

The most significant climate processes since the middle to late Pliocene are the glacial and interglacial cycles. The present interglacial period has lasted about 11,700 years. Shaped by orbital variations, responses such as the rise and fall of continental ice sheets and significant sea-level changes helped create the climate. Other changes, including Heinrich events, Dansgaard–Oeschger events and the Younger Dryas, however, illustrate how glacial variations may also influence climate without the orbital forcing. Glaciers leave behind moraines that contain a wealth of material—including organic matter, quartz, and potassium that may be dated—recording the periods in which a glacier advanced and retreated. Similarly, by tephrochronological techniques, the lack of glacier cover can be identified by the presence of soil or volcanic tephra horizons whose date of deposit may also be ascertained.

Vegetation

A change in the type, distribution and coverage of vegetation may occur given a change in the climate; this much is obvious. In any given scenario, a mild change in climate may result in increased precipitation and warmth, resulting in improved plant growth and the subsequent sequestration of airborne CO_2. Larger, faster or more radical changes, however, may result in vegetation stress, rapid plant loss and desertification in certain circumstances.

Ice cores

Analysis of ice in a core drilled from a ice sheet such as the Antarctic ice sheet, can be used to show a link between temperature and global sea level variations.

The air trapped in bubbles in the ice can also reveal the CO_2 variations of the atmosphere from the distant past, well before modern environmental influences. The study of these ice cores has been a significant indicator of the changes in CO_2 over many millennia, and continues to provide valuable information about the differences between ancient and modern atmospheric conditions.

Dendroclimatology

Dendroclimatology is the analysis of tree ring growth patterns to determine past climate variations. Wide and thick rings indicate a fertile, well-watered growing period, whilst thin, narrow rings indicate a time of lower rainfall and less-than-ideal growing conditions.

Pollen analysis

Palynology is the study of contemporary and fossil palynomorphs, including pollen. Palynology is used to infer the geographical distribution of plant species, which vary under different climate conditions. Different groups of plants have pollen with distinctive shapes and surface textures, and since the outer surface of pollen is composed of a very resilient material, they resist decay.

Changes in the type of pollen found in different layers of sediment in lakes, bogs, or river deltas indicate changes in plant communities. These changes are often a sign of a changing climate. As an example, palynological studies have been used to track changing vegetation patterns throughout the Quaternary glaciations and especially since the last glacial maximum.

Insects

Remains of beetles are common in freshwater and land sediments. Different species of beetles tend to be found under different climatic conditions. Given the extensive lineage of beetles whose genetic makeup has not altered significantly over the millennia, knowledge of the present climatic range of the different species, and the age of the sediments in which remains are found, past climatic conditions may be inferred.

Sea level change

Global sea level change for much of the last century has generally been estimated using tide gauge measurements collated over long periods of time to give a long-term average. More recently, altimeter measurements — in combination with accurately determined satellite orbits — have provided an improved measurement of global sea level change. To measure sea levels prior to instrumental measurements, scientists have dated coral reefs that grow near the surface of the ocean, coastal sediments, marine terraces, ooids in limestones, and nearshore archaeological remains. The predominant dating methods used are uranium series and radiocarbon, with cosmogenic radionuclides being sometimes used to date terraces that have experienced relative sea level fall.

Index

P

R

S

T

W

Y

Z